4D CAD and Visualization in Construction

4D CAD and Visualization in Construction: Developments and Applications

Raja R.A. Issa
Ian Flood
William J. O'Brien
University of Florida, Gainesville, USA

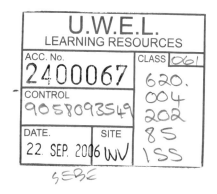

A.A. BALKEMA PUBLISHERS   LISSE/ABINGDON/EXTON (PA)/TOKYO

Library of Congress Cataloging-in-Publication Data

(applied for)

Cover design: Studio Jan de Boer, Amsterdam, the Netherlands
Typesetting: Charon Tec Pvt. Ltd, Chennai, India
Printed in the Netherlands by Grafisch Produktiebedrijf Gorter, Steenwijk

© 2003 Swets & Zeitlinger B.V., Lisse

Published by: A.A. Balkema Publishers, a member of Swets & Zeitlinger Publishers
www.balkema.nl and www.szp.swets.nl

ISBN 90 5809 354 9

Table of contents

Foreword

The final frontier of the application of information technology in construction is the job site. And perhaps there is no stronger technological link between the job site and the design office than the practice of 4D CAD, and for good reason. Implementing a 4D CAD system cuts to the very heart of issues that mankind has struggled with for centuries: linking space and time; turning concept into reality; ownership of knowledge; effective communication between business partners.

Models have been used for centuries to explain what is to be built, and the promise of this new technology is exciting. Never before in history has man, without ever turning a single shovel of earth or driving a single nail, been able to virtually construct a bridge or building one piece at a time and link each step to a corresponding step on a schedule.

Nonetheless, entrenched ways of working do not change overnight. At the time of writing, the flurry of activity with dot.com investments and startups has slowed to a trickle. It comes as no surprise to many industry observers that many of the quick fixes with technology promised by these startups failed to take hold. What did take place, however, was a demonstration that the construction industry continues to be ripe and ready for new technologies that will provide value.

A problem still plaguing the industry is that many in it are not in sync with the technology. A growing body of workers is well versed in the ways of the computer, but novice at best in the ways of construction. Juxtapose that with a still significantly sized body of industry veterans that have yet to embrace technology. Arguments can be made in support of delays to widespread adoption of 4D CAD are that the technology is not evolved enough for the people, or that the people are not evolved enough to use the technology—both viewpoints are valid. Those assertions are, at a deeper level, the continuing story of mankind's relationship with his tools.

Using contractual methods such as design-build to force parties to work together, rather than technological tools, has met with some success. On the horizon, it could also be a driving force to help the growth of 4D CAD. Superficially, a simple schedule and a CAD model may qualify as 4D CAD. Yet to make the practice truly widespread, deeper issues such as trade sequencing, production planning and dynamic cost and resource planning need to be addressed.

No industry needs 4D CAD more than construction. Like NASA's Space Program, and the developments borne from it that ultimately benefited mankind as a whole, 4D CAD is similarly poised to have a major impact on any industry, including construction, that struggles to simultaneously manage the scheduled creation of objects, be they airplanes, airports or air cleaners.

When designers have a better grasp of scheduling and buildability issues facing contractors, projects will be better. Similarly, when contractors are more involved earlier in the process, buildability issues will be dealt with on screen or on paper, instead of in situ. And without the disputes, disagreements and the rampant adversarial relationship that so often plagues construction projects today, designers and constructors can better concentrate on what they want most, which is ultimately to build.

Matthew Phair
Senior Editor
Engineering News-Record

BENEFITS OF 3D AND 4D MODELS FOR FACILITY MANAGERS AND AEC SERVICE PROVIDERS

Martin Fischer[1], John Haymaker[2], Kathleen Liston[2]

[1] *Civil and Environmental Engineering and (by courtesy) Computer Science, Stanford, CA, USA*
[2] *Civil and Environmental Engineering, Stanford, CA, USA*

Abstract

The first part of this paper presents an extensive list of benefits users of 4D models have realized and illustrates the benefits with specific examples from actual uses on a variety of projects. It illustrates how current business practices and project delivery approaches allow or do not allow facility owners to reap these benefits. All owners and AEC service providers (designers, general contractors, subcontractors) who have used 4D models to assist in understanding, analyzing and communicating a design and construction schedule have reported benefits from the use of these models. Owners have used 4D models to plan the construction of facilities that require significant phasing prior to contract award to verify the overall constructibility of a proposed design given the project timeline and available space. General contractors have used 4D models for overall and for detailed construction planning, to communicate scope and schedule information effectively to subcontractors and other parties, and to test the constructibility of the design and the executability of the schedule prior to committing resources to the field. The second part of the paper describes in detail the application of 4D models for construction scheduling and constructibility analysis on the Walt Disney Concert Hall in Los Angeles. It discusses the reasons for the use of 4D models on the project and details the technical challenges the 4D modeler had to overcome. Specific examples of the impact of the 4D model on the schedule are also shown.

Keywords: 4D modeling, construction planning, case studies, benefits

INTRODUCTION

Traditional construction planning tools, such as bar charts and network diagrams, do not represent and communicate the spatial and temporal, or 4D, aspects of construction schedules effectively. Consequently, they do not allow project managers to create schedule alternatives rapidly to find the best way to build a particular design. Extending the traditional planning tools, visual 4D models combine 3D CAD models with construction activities to display the progression of construction over time. 4D models combine 3D CAD models with the project timeline (Cleveland, 1989). Systems linking 3D CAD models with schedule and other project information started to be developed in the mid-eighties (Kahan & Madrid, 1987; Atkins, 1988). Experience on many different types of projects (simple to complex, new to retrofit) has shown that combining scope and schedule information in one visual model is a powerful communication and collaboration tool for technical and non-technical stakeholders (Williams, 1996; Retik, 1997; Edwards & Bing, 1999).

The 4D research team at Stanford University has tested the usefulness of visual 4D models in planning the construction of a hospital, the roof of a university building, a small commercial building, a Frank Gehry designed museum, a theme park, and a Frank Gehry designed concert hall. These cases have shown that more project stakeholders can understand a construction schedule more quickly and completely with 4D visualizations than with the traditional construction management tools. Since they understand the scope and schedule of a project better, the stakeholders can then provide input to the scope and schedule and the important interrelationships, and help improve the project design and schedule. We and other 3D and 4D practitioners found that project managers using 4D models are more likely to allocate resources (e.g. design time, client review time, management attention, construction crews) more effectively than those who do not use 4D models. Danhier et al. (1994) came to similar conclusions in their application of 4D models to the replacement of steam generators.

3D CAD is often seen mainly as a design tool. It should also be seen as a construction tool, since a detailed 3D CAD model mirrors the completed project in the computer. It affords a project team the opportunity to practice or rehearse the construction of a unique artifact virtually before building it in reality. Project teams need to decide what problems they want to resolve through the use of 3D and 4D models. The resulting purpose of the 3D and 4D modeling effort has implications on who needs to be involved in the modeling effort. Should the models help answer questions to overall site logistics, flow of work, or access to various parts of the project at various times? Or should the models help answer questions about the specific sequence of work for a group of subcontractors, the laydown spaces needed for particular activities, or the distance in time and space between succeeding work?

For example, effective use of 3D and 4D CAD as a detailed construction tool has implications on the project delivery process, the output or deliverables of various parties, and the processes and organization of projects. If the 3D model is to mirror the real project in detail, the same organizations that build the project should build the model because they will have the biggest stake in the accuracy of the information in the model. It is also unrealistic to expect that a group of designers and modelers has all the expertise about construction details necessary for a detailed 3D model. The experience of the 4D research group at the Center for Integrated Facility Engineering (CIFE) at Stanford University shows that including at least key subcontractors as design-build firms from the beginning of a project makes detailed 3D modeling more efficient and effective than including them later (Staub et al., 1999).

It is difficult for designers to know to what level of detail they should model a particular part of a project, since they often do not benefit directly from accurate detailed 3D models that clearly show what needs to be built. We have found that the subcontractors, however, are very interested in having accurate, reliable, and well-coordinated detailed design information because they can leverage that information in material procurement and management, and in planning and scheduling. Building a 3D CAD model in this way leaves accountability for the correctness of the information in the 3D model with the firms who are best equipped to leverage the investment in building 3D models. Designers remain in charge of the overall design concept, and subcontractors can focus on streamlining the production of their part of a project.

A detailed and well-coordinated 3D CAD model allows firms to prefabricate directly from the model and improves material management. In this way, 3D CAD models enable project managers to allocate and use material resources more efficiently. 4D models extend the usefulness of design information to the construction planning and construction phases. If 4D models are built during the design phase they can help provide constructibility feedback to the design team, and they can also help set priorities for design work so that the necessary material procurement and crew planning information for construction are available in a timely manner. In this way, 4D CAD models help project managers to manage the flow of work and the allocation of crews and space on construction sites better (Vaugn, 1996). The next sections introduce benefits companies using 3D and 4D CAD models have realized in more detail. The later sections in this paper give specific examples of uses of 4D models and describe the parties that participated in the modeling efforts and discuss the level of detail they found useful.

Table 1 lists benefits of 3D and 4D models realized by companies using such models for design and construction. Engineers and managers from owner, design, and construction firms reported them at a workshop on the use of 3D and 4D models hosted by Walt Disney Imagineering (WDI) and CIFE in May 1999.

The column on the left lists specific benefits users have reported, the two middle columns show who realizes the benefit and who has most control and influence over the information in the 4D model necessary to realize the benefit. The column on the right shows whether the beneficiary matches the controlling party. Y means that it does, N means that it does not and S means "somewhat", i.e. the primary beneficiary has significant control over the information in the 4D model necessary to realize the benefit, but other parties have some influence as well. Rows where the party that realizes the benefit does not match with the party controlling

Table 1.   Benefits of 4D models for owners, designers, general contractors and subcontractors*.

Benefit	Realize	Control	Influence	R = C?
Reduce design time	D	D	O	S
Reduce design effort	D	D	O	S
Speed up evaluation of design	D	D	O	S
Reduce time needed to model an alternative	D	D		Y
Improve evaluation of design (functional sensitivity analysis)	D	D		Y
Share work around the world (model-centric project teams)	D	D		Y
Eliminate design production work (CD)	**D**	**O**		**N**
Increase and improve information available for early decision-making	DO	DO	SGC	S
Reduce project management costs	GC	GC	SDO	S
Improve evaluation of schedule	GCS	GCS	D	S
Reduce number of change orders	**O**	**D**	**SGC**	**N**
Increase number of alternatives studied	**O**	**D**		**N**
Increase number of project stakeholders who clearly understand the project and who are able to provide input	**O**	**D**		**N**
Shorten (simplify, streamline) permitting time and effort	**O**	**D**		**N**
Increase concurrency of design and construction	O	DGCS		S
Reduce interest costs	**O**	**GC**	**SD**	**N**
Reduce time to make a decision	O	O	D	S
Obtain management decision, funding	O	O	D	S
Reduce life-cycle costs	O	O	D	S
Maximize value to owner	**O**	**SGCD**		**N**
Increase productivity of crews	S	S	GCDO	S
Reduce wasted materials during construction	S	S	GCDO	S
Reduce rework	S	S	GCDO	S
Create complete information to build from	**S**	**SD**		**N**
Improve (verify, check) constructibility	**SGC**	**D**		**N**
Verify consideration of site constraints in design and schedule (sight lines, access, …)	**SGC**	**D**		**N**
Avoid (minimize, eliminate) interferences on site	**SGC**	**D**		**N**
Maximize off-site work (prefabrication)	**SGC**	**D**		**N**
Increase schedule reliability	SGC	SGC	D	S
Verify executability of GC and sub-schedules	SGC	SGC	D	S
Shorten construction period	SGC	SGC	DO	S
Speed up evaluation of schedule	SGC	SGC	O	S
Increase site safety	SGC	SGC		Y
Minimize in-process time in supply chain	SGC	SGC		Y
Shorten site layout/surveying time	SGC	SGC		Y
Improve site layout accuracy	SGC	SGC		Y
Reduce RFIs	**SGCD**	**D**		**N**
Improve portability of design	**SGCD**	**D**		**N**
Shorten design and construction period	SGCD	SGCD	O	S
Improve learning and feedback from project to project	**SGCDO**	**O**	**D**	**N**
Improve effectiveness of communication	**SGCDO**	**O**	**SGCD**	**N**
Bring new team members up to speed quickly	SGCDO	SGCDO		
Coordinate owner, GC and sub-schedules	SGCO	SGCO		Y

* Keys: O = owner, D = designer, GC = general contractor, S = subcontractor.

or generating the information are shown in bold. The assignments to who controls the data and who realizes a benefit assume a traditional project delivery process. Table 1 shows that many benefits that potentially translate into significant time and cost savings are unlikely to be realized with a traditional project organization because the party benefiting from the use of 3D and 4D models is not in control of the information necessary to realize the benefit.

Benefits for designers
It is commonly understood that a design documented with a 3D CAD model will most likely have fewer errors and coordination issues because the construction of the model by multiple designers forces and allows them to reconcile inconsistencies. Evaluation of a design in 3D is also faster than with 2D drawings because reviewers can more quickly understand the scope and status of the design. Workshop participants who have been using 3D CAD models for several years reported that, after an initial learning curve, the overall design effort and design time is less than with a process using 2D drawings. They use 3D even when the client asks for 2D drawings because design revisions are faster and need to be done only once (instead of updating plans, sections, elevations and details). A further benefit is the potential to eliminate construction documents. Most participants saw little value in most 2D construction documents currently produced by design firms. On many projects subcontractors complete a new set of shop drawings anyway, and in some cases subcontractors fabricate parts directly from the 3D CAD model with numerically-controlled machines. 2D construction documents and shop drawings appear to be rather useless on a project where the design is documented and shared with a detailed 3D CAD model. Designers involved in projects that used 3D models from design through construction reported that they saw an increased coordination effort during the design phase of the project followed by fewer requests for information during construction. Hence, designers were able to focus on the phase of the project they enjoy most.

Benefits for owners
Owners are, of course, the ultimate beneficiaries of better performance by designers and builders from the use of 3D and 4D models. The workshop participants noted, however, that owners can use 3D and 4D models themselves to speed up and improve decision-making and to involve many more stakeholders than traditionally possible. For example, WDI was able to get the input from about 400 stakeholders during the two-month pre-bid design and construction schedule review for the Paradise Pier portion of Disney's California Adventure. They were holding meetings with groups of eight to ten people at a time in their Computer-Assisted Virtual Environment (CAVE). The groups could interactively review the proposed design and construction schedule from any perspective and quickly understand the design, schedule, and corresponding constraints (Fischer et al., 2001).

Benefits for builders

All participants at the workshop who have labor risk on site reported that detailed 3D and 4D models greatly increase the productivity of crews and help eliminate wasted materials and resources. Even if all the other project team members are working with 2D drawings many subcontractors still elect to build a 3D model for their scope of work and for the related scope of work. If all the information is readily available, they can build the 3D and 4D CAD models to verify that no interferences exist and that they have all the information and materials available for construction. If the information to build a detailed 3D CAD model is not available it is far cheaper for an engineer in the office to figure out what exactly needs to be built than for a crew in the field. The 3D models also support automated quantity takeoff for material procurement to ensure that each crew has the appropriate amounts and types of materials for a given day or week's work.

General Contractors (GCs) and subcontractors benefit from smooth, safe, and productive site operations, since that contributes to the shortest and most economical construction period. If built from subcontractor and GC schedules 4D models help the construction team coordinate the flow of work and space use on site. Contractors usually produce phasing drawings for a project. Typically, they are done manually, which makes it difficult to communicate them to all the interested and affected parties in a timely manner when they can still be improved economically. It also makes updating of the phasing plans a chore. Furthermore, they are only produced in 2D and for a few snapshots in time, which makes it more likely that a potential interference between trades gets overlooked. Combining 3D models with schedules automatically produces 3D phasing drawings at the daily, weekly, or monthly level depending on the level of detail in the schedule and the 3D CAD model. Contractors can easily see who is working where on what and how the work proceeds over time and through the site. 3D phasing drawings automatically reflect schedule updates.

In summary, all workshop participants found that 4D models communicate schedules much more effectively than the abstract bar charts used on most projects, which, in turn enables the benefits listed in Table 1. Songer et al. (1998) came to similar conclusions in their study.

EXAMPLES OF APPLICATIONS OF 4D MODELS AND CORRESPONDING BENEFITS

Members of the 4D research group at CIFE have supported construction project teams in applying 4D models on their projects since 1993 and have used the insights gained from applying the 4D models in real world settings to drive research efforts. Fischer & Aalami (1996), Akinci & Fischer (1998), and Fischer et al. (1998) give examples of the use of observations on construction projects to formulate research questions, help formalize specific knowledge, and test research prototypes. This section briefly discusses these applications of 4D modeling and summarizes the corresponding benefits. In our experience, 4D models offer benefits on simple and complex projects, on new construction and on retrofit projects, and at the detailed nuts and bolt level as well as for overall project phasing. Unless noted otherwise, we used the Bentley Schedule Simulator or an earlier version of this program to combine 3D CAD and schedule information. Most projects were modeled in 3D with AutoCAD, and the schedule information was mostly in Primavera's P3 tool, although MS Project has also been used. These brief descriptions are followed by an in-depth case study.

1993–95: RECONSTRUCTION OF THE SAN MATEO COUNTY HEALTH CENTER

Together with the GC, Dillingham Construction Company, CIFE researchers build 3D and 4D models to coordinate the overall master plan so that the five-year construction period interfered as little as possible with the operation of the hospital. The 4D model coordinated owner relocation and operations schedules with construction schedules, facilitated client input, eased relationships with the community, and became, according to the hospital director, the best fund-raising tool (Collier & Fischer, 1996). It allowed, for example, verification that hospital staff could always reach all parts of the hospital from any other part without leaving the hospital. A more detailed 4D model was used to verify the constructibility of the central utilities plant and to make sure that the design and schedule information were complete and well coordinated. The 4D model for the US $100 m, 320,000 sf project was built from about 25,000 3D CAD elements and 500 activities in about 1,000 hours. The 4D model to study the overall phasing of the project consisted essentially of the main architectural components of the project: walls, windows, doors, columns, slabs, roofs. The 4D model for the central utilities plant was more detailed and included the foundations, some of the stud walls, the equipment platforms and equipment and the major mechanical ductwork. Collier & Fischer (1995) provide detailed information on the 4D modeling effort required for this project. The 4D modeling effort was carried out in parallel to the GCs construction planning efforts. Hence the 4D modeling effort served to confirm the project manager's thinking about his approach to construction. It also helped the project manager communicate how hospital operations and construction were going to coexist in close proximity for the scheduled five-year duration of the project.

The paragraphs below explain how 4D modeling helped improve the construction schedule for the San Mateo County Health Center. Figures 1 to 8 show eight stages in the reconstruction of the San Mateo County Health Center project over the planned five-year construction period from May 1994 to June 1999. The first

Figure 1.   May 1994. Site with existing buildings, Main Hospital at left, Clinics Building at right foreground, East Wing in between, Existing Utilities plant in middle, Aids Clinics at top right.

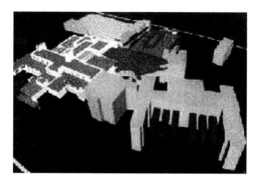

Figure 2.   March 1995. As work on the new Central Utilities Plant proceeds as the critical activity, construction on the first half of the North Addition begins. Trailers for temporary office space are installed next to the Clinics building. The first portion of the Central Hub is under construction (shown in green in the center of the snapshot). The hospital operations are linked through the existing East Wing (shown in gray in the center of the snapshot).

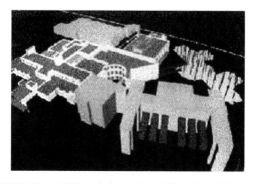

Figure 3.   October 1995. Central Plant is completed, North Addition interior work is being completed. Aids Clinics is now in trailers. Aids Clinics is demolished and work has begun on new Nursing Wing.

Figure 4.   January 1996. Nursing Wing exterior shell begins construction as critical path activity. East Wing functions are moved to North Addition. Connector is built to link first half of Central Hub to Clinics Building. The first portion of the Hub is finished and can now serve as a link between departments. This makes it possible to demolish the East Wing (shown in green in the center of the snapshot) to make room for the second half of the Central Hub.

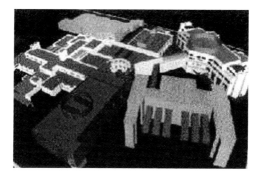

Figure 5.   July 1996. East Wing has been demolished and new Clinics Building and second half of North Addition are under construction in its place. Nursing wing work has moved to the interior.

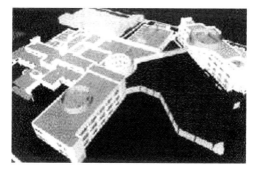

Figure 6.   June 1997. Old Clinics Building has been demolished and cleared away. Construction of the new Diagnostics and Treatment Building has begun.

Figure 7.   December 1998. Remodeling of existing hospital structure begins. Entire third floor and ancillary wings are removed.

Figure 8.   June 1999. The new San Mateo County Health Center.

snapshot from the 4D model (May 1994) shows a view of the hospital prior to construction (Fig. 1), and the last snapshot (June 1999) shows the 3D CAD model of the reconstructed hospital (Fig. 8). As can be seen from comparing these two models, the transformation of the hospital during the reconstruction period is very dramatic. Constructing new parts and renovating other parts of the hospital without interrupting hospital operations was a challenging task.

Originally, the GC showed the construction period with a bar chart schedule based on a critical path network. As can be imagined by just looking at the "before" and "after" models, such an abstract representation of the flow and sequence of construction fails to uncover potential time-space conflicts between construction and operations. The six snapshots of the planned progress of the construction (Figs 2 to 7) taken from the simulation of the construction schedule show the relationship between construction activities and hospital operations more clearly. It is easy to create as many snapshots as desired. In these snapshots, building activities are shown in green (non-critical activities) and red (critical activities). Parts of the facility that are not under construction, i.e. where construction has not yet started or where construction has been completed, are shown in gray and yellow (the original colors in the 3D model).

The following example illustrates the usefulness of 4D models to support master planning. We have also tested the usefulness of 4D models for more detailed planning and found them to be extremely helpful for the coordination of contractors and for constructibility improvements (see the other case studies in this paper). The 4D model, from which the snapshots are taken, took about four person-months to build.

Instead of focusing on all the activities that are shown in the snapshots, we would like to draw the reader's attention to an improvement to the interaction between construction and operations the 4D modeling effort helped make (the snapshots show the improved version). Initially, the Central Hub (the building with the round yard in the center of the hospital), connecting all parts of the hospital, was to be constructed as one building at one time. However, if that had happened (as was confirmed with the first 4D simulation), construction would have cut hospital operations in half. It would have been necessary to put patients on gurneys, wheel them out the door and around the block to bring them from their rooms to radiology services. This was, of course, not acceptable for the hospital staff, and the designers and construction managers had to find a solution to maintain uninterrupted links within the hospital between all hospital departments at all times. They decided to cut the Central Hub in half, add a seismic joint in the middle, and build one section of the Hub in the early phases of the project, as seen in the March 1995 snapshot (Fig. 2).

1995: ROOF FOR HAAS SCHOOL OF BUSINESS, UC BERKELEY

The 4D modeling effort on the San Mateo Health Center project was done as part of the early construction planning phase and informed the project management team about potential problems and opportunities for improvement. In contrast, we completed a small study of the applicability of 4D models to day-by-day subcontractor coordination after the work had been done. The advantage was that we knew why and in what way the construction of the roof had not been as efficient as possible. Misunderstandings between the architect, the GC and the roofing, stucco, and sheet metal subcontractors led to extra cost of about US $200,000 due to low productivity and rework. Together with the roofing subcontractor we developed 4D CAD models of the various design solutions and several construction sequences using keyframes produced with 3D Studio in less than 40 hours. The 4D model included all the parts including the main assembly pieces that needed to be installed on the roof. The model clearly showed the challenges and tradeoffs of the various design and schedule proposals and would have been helpful for the contractors to understand each others' constraints. Fröhlich et al., 1997 show snapshots from this 4D model.

1997–99: SEQUUS PHARMACEUTICALS PILOT PLANT IN MENLO PARK

The 4D model for this biotech project coordinated the mechanical, electrical, and piping (MEP) contractors' day-by-day work. As a result there were no field interferences, no rework, higher productivity, only one contractor-initiated change order, no cost growth during construction, and 60% fewer requests for information than expected for this type of project (Staub et al., 1999). The GC also used the 3D model for automated quantity takeoff. The Stanford 4D group built the 4D models on this project with input from the GC and from the MEP subcontractors. The model was very detailed, including all components that needed to be installed for the scope of work of the MEP subcontractors down to 50 mm (2 inch) piping.

1998: McWHINNEY OFFICE BUILDING, COLORADO

The 4D model for this small commercial project allowed junior engineers to improve a CPM schedule developed by the project manager and superintendent or the GC (Koo & Fischer, 2000). The improvements could have saved about two weeks in project duration. This study was also done after construction was completed. It demonstrated that 4D models have the potential to make junior engineers productive contributors to getting a project built.

1998: EXPERIENCE MUSIC PROJECT (EMP), SEATTLE

The 4D models for this project with extremely complex geometry visualized various schedule versions so that the owner representative, architect, and GC could more easily understand the repercussions of, for example, delaying a decision. 4D models also showed detailed construction sequences. Although the architect, Frank O. Gehry and Associates (FOGA), made the 3D models available to the GC, the 4D modeler needed to add significant construction detail to the 3D model to generate a realistic 4D visualization (Fischer et al., 1998).

1998–99: PARADISE PIER, DISNEY CALIFORNIA ADVENTURE

A 4D model including staging and laydown areas allowed the owner's construction planning team to verify that the project timeline requested in the bid documents was aggressive but realistic. The 4D model became part of bid documents. The owner, WDI used the 4D models in pre-bid meetings with the invited GCs to explain the scope and challenges of the project. The winning bid came in slightly under WDI's budget and proposed a schedule that was two months shorter. Throughout the owner's construction planning effort, the owner used the 4D models on desktops and in a CAVE to support design and schedule reviews. WDI further leveraged its investment into the 3D model developed for the 4D model to check 3D sight lines and to simulate the rides. On this project, WDI and CIFE researchers collaborated to develop a prototype 4D tool that emphasizes ease of use and interactivity. The prototype allowed planners to work with the 4D model at several levels of detail and make changes to the 3D model and schedule in the 4D environment (Schwegler et al., 2000).

WALT DISNEY CONCERT HALL

The rest of the paper describes our most recent involvement in 4D modeling efforts on an ongoing construction project.

THE PROJECT, PARTICIPANTS AND MOTIVATION FOR 4D MODELING

The Walt Disney Concert Hall (WDCH), designed by FOGA, is the new 2,300 seat home of the Los Angeles Philharmonic Orchestra. Located in downtown Los Angeles, the US $240 m project incorporates complex architectural, structural, and acoustical requirements in a tight one-city-block site. The project is scheduled for completion in early 2003. Figure 9 shows a photo of the front entrance to the WDCH.

Figure 9.   Photo of the physical model of the Disney Concert Hall.

The architectural process undertaken by FOGA provides(opportunities and challenges for the construction of 4D models to assist in the construction planning process.) FOGA's design process yields a highly developed 3D CAD product model, which is used extensively for dimensional control and fabrication in the construction process. This product model and the process model contained in the construction schedule prepared by M.A. Mortenson Company, the GC, with input from many subcontractors, provide the necessary elements to begin construction of the 4D model. The GC used the 3D and 4D models as communication tools to share project information with all project participants including architects, engineers, the GC, subcontractors, and the owner. John Haymaker from the 4D research team at Stanford University worked on site to help build the 4D models discussed below and to introduce the GC and key subcontractors to the 4D modeling process. He used the prototype 4D modeling software developed through the collaboration of the Research and Development group at WDI and researchers in the 4D CAD research group at the CIFE at Stanford University (Fischer et al., 2001).

The complex project and a tight site made precise coordination of construction activities a very high priority. M.A. Mortenson saw the use of 4D visualization of the construction process as a valuable tool for accomplishing four project objectives:

Schedule creation: 4D models help visualize schedule constraints and opportunities for schedule improvements through resequencing of activities or reallocation of work space.

Schedule analysis: 4D models help analyze the schedule and visualize conflicts that are not apparent in the Gantt charts and CPM diagrams.

Communication: Many participants join the project in midstream, and it is critical to bring new participants up to speed quickly.

Team building: The GC's project superintendent, Greg Knutson, felt strongly that it was very important to construct a team atmosphere, where people solve problems together. He realized that a shared, visual model to externalize and share project issues was a valuable team building tool.

The following section details the project information that was available at the beginning of the 4D process. Subsequently the process undertaken to construct the 4D models and describe the 4D models constructed for the project is examined. We have also described some of the issues and challenges encountered in constructing the models. The final discussion focuses on how the GC used the models to accomplish the objectives.

AVAILABLE ELECTRONIC INFORMATION

The interest in constructing the 4D models emerged in early 2000, as the GC mobilized to the site. At this point, the architect had already developed most of the

3D geometry, and the GC's construction schedule had about 4,000 activities. This section describes the format and level of detail of the project information at the beginning of the 4D modeling process.

AVAILABLE 3D GEOMETRY

The architect constructed the 3D models with CATIA. There are at least two reasons for the use of CATIA as the 3D modeling software. First, FOGA develops very complex geometry and considers the nature of the curves generated to be integral to the architectural design. CATIA uses NURBS-based curves and surfaces, which describe the curves mathematically, and therefore maintain a high level of accuracy. More traditional CAD packages for the AEC industry do not use NURBS, instead approximating the curves and therefore loosing the level of accuracy desired by FOGA. The second motivation for using CATIA is that the software handles very large, complex models. As described below, the architect modeled a great deal of the project in 3D, and the shear amount of information would overwhelm traditional AEC CAD packages.

To reduce complexity, FOGA divides the 3D model into sub-models. First, FOGA divides the project geographically into "building elements," as shown in Figure 10.

Each building element is then further divided into models reflecting different significant building systems. Figures 11 to 17 show the different models available

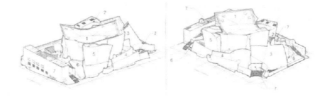

Figure 10.   WDCH broken down by building element.

Figure 11.   Surface models for all building elements.

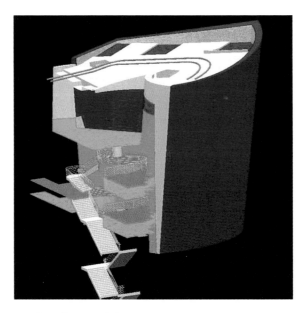

Figure 12.   Element 2 surface model.

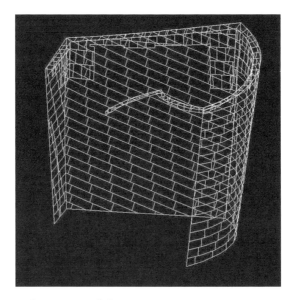

Figure 13.   Element 2 pattern model.

for building element 2. Figure 11 shows all of the building elements' surface models incorporated into one view. Figure 12 shows the surface model for element 2. The surface model contains everything that can be seen, from plaster, to glazing, to carpet, to wood panneling, etc. Figure 13 shows a pattern model. A pattern model describes any pattern in an element that is relevant for architecture or construction.

Figure 14. Concrete model for element 2.

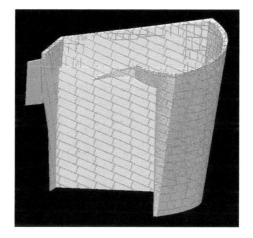

Figure 15. Air and water barrier model for element 2.

Figure 13 shows the pattern of the stainless steel panels for the exterior of element 2. Figure 14 shows the concrete model, which models the structural and architectural concrete surfaces. Figure 15 shows an example of an air and water barrier model. The air and water barrier model defines the surface in space where the water-proofing systems should be placcd. Figure 16 shows the structural wireframe model. This model defines a wire for each piece of steel in the building. The wire can symbolize centerline, top of steel, or bottom of steel. The steel detailer and the steel fabricator use this wire model as input and place the proper size member with each wire. The detailers detail all the connections in X-Steel or other detail-ing packages in 3D. The resulting detailed steel model, shown in Figure 17, is then re-imported into the CATIA model.

Figure 16.   Element 2 steel wireframe model.

Figure 17.   Detailed steel model.

Each 3D model consists of layers reflecting different sub-systems. Table 2 shows a partial listing of the layers. These layers are helpful for 4D modeling because they isolate certain scope information in the 3D model, which facilitates the identification of the appropriate geometric elements for a particular activity. However, frequently the layering organization is different from the organization of

Table 2. A portion of the layer list.

Layer No.	CATIA layer contents	Layer No.	CATIA layer contents	Layer No.
	General project data (1 Thru 10)		Stone (46 Thru 55)	
1	Project grid	46	Vertical stone cladding	86
2	Column grid lines	47	Sloped stone cladding	87
3	Property line	48	Stone coping	88
4	Vacation envelope	49	Stone paving	89
5	Project reference geometry	50	Stone base	90
6	Project workpoints	51	Decomposed granite	91
7	CATIA construction geometry	52	Not used	92
8	Existing construction	53	Not used	93
		54	Not used	
	Glazing assemblies (11 Thru 25)	55	Not used	
11	Skylight glazing		Roof asssemblies (56 Thru 65)	96
12	Sloped glazing			97
13	Vertical glazing	56	Roof membrane Type 1	98
14	Mullion wireframe (Center Line Mullion)	57	Roof membrane Type 2	99
15	Mullion	58	Roof hatch	100
16	Metal closure Trim	59	Roof drain	101
17	Metal closure Panels	60	Stainless steel gutter	102
18	Metal gutter	61	Expansion joint assembly	103
19	Metal flashing	62	Roof davit pedestal	104
20	Glazing anchor assembly	63	Roof assembly Type 3	105
21	Glazing boundary	64	Not used	106
		65	Not used	107
	Metal panel assemblies (26 Thru 45)			108
			Miscellaneous exterior assemblies	109
26	Metal panel assembly condition Type 1			110
27	Metal panel assembly condition Type 2	66	Not used	
28	Metal panel assembly condition Type 3	67	Metal grill	
29	Metal panel assembly condition Type 4	68	Metal grating	
30	Metal panel assembly condition Type 5	69	Building maintenance equipment	126
31	Metal panel assembly condition Type 6	70	Building maintenance track	127
32	Metal panel assembly air and water barrier	71	Stain less steel clad door	128

the schedule, and the 4D modeler needs to reorganize the geometric information for the 4D model to fit the schedule organization (Fischer et al., 1998).

AVAILABLE SCHEDULE INFORMATION

The GC created the construction schedule with Primavera's P3™ software. At the start of the 4D modeling process in March 2000, the schedule contained about 4,000 activities. By Fall 2000, the schedule consisted of approximately 7,200 activities. The schedule divides the 3D project geometry into chunks that are relevant to an activity. Figure 18 shows the breakdown key for the activity ID in the schedule. Activities are identified by building element, floor, area, and subarea, then by phase, system, component, and action. However, some activities do not fit easily into this breakdown. For example, steel installers like to break the steel into manageable chunks, called sequences, which are a grouping of steel that is self-supporting and can be erected in a reasonable amount of time. These sequences often span more than one building element, or cover more than one floor. Even though it was useful to have one main way to organize the schedule (as shown in Fig. 18), many methods for decomposing the geometry and linking a scope of work to an activity are required to suit different types of work. Figure 19 shows

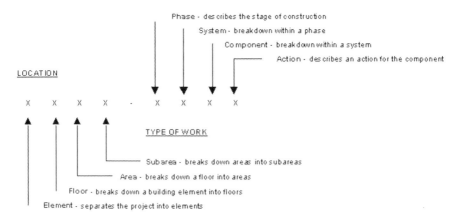

Figure 18.   Activity code key for defining activities and relating them to the 3D model.

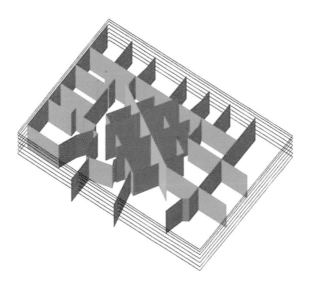

Figure 19.   Organization of 3D model into levels, sequences and thirds.

the project broken into levels (red) and sequences (green). Figure 19 also shows the main potion of the Concert Hall broken into thirds (blue) as the GC organized some of the work in the main hall in this way.

4D MODELING PROCESS AND 4D MODELS

Figure 20 maps the process for constructing the 4D models from the project geometry and schedule and shows the file formats used to translate between computer programs. Rhino3D™ proved to be very useful to import the NURBS-based geometry from CATIA, add names to the geometry, break up the geometry into relevant configurations for the respective activities, and convert the geometries to VRML. Named geometrical elements allow a 4D modeler to match geometry names to activity names quickly.

We built four 4D models for the project. Figures 21 to 24 show a screen shot from each of these models.

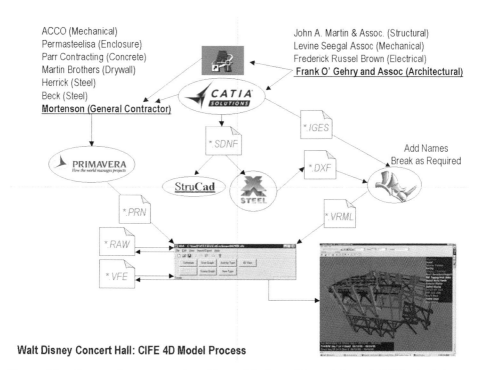

Walt Disney Concert Hall: CIFE 4D Model Process

Figure 20.   Process for constructing 4D models from 3D models and CPM schedules.

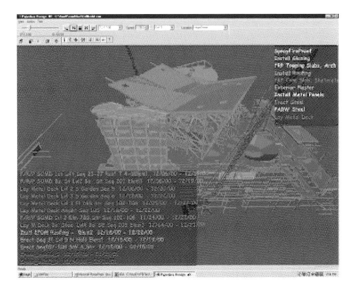

Figure 21.   Steel, Concrete, and Exterior Enclosure model. This 4D model examines the overall sequencing for the major structural and enclosure activities. It shows the sequencing of steel and of structural and architectural concrete. It includes metal decking, roofing, glazing, and enclosure systems, such as metal cladding assemblies including secondary steel supports. Statistics: Number of 3D components: 340; Number of polygons: 515,000; Number of activities: 512.

Figure 22.   Element 2 model. This 4D model goes into more detail for building element 2. It includes the interior work. The model includes interior stairs, elevators, fireproofing, and finishing systems. It shows mechanical and electrical activities by highlighting the floor slabs in the area of work. Statistics: Number of 3D components: 105; Number of polygons: 85,000; Number of activities: 185.

Figure 23. Interior hall model. The interior of the Concert Hall is a highly congested and complex space. All of the interior activities are squarely on the critical path. The model includes all the activities affecting this space: structural steel, concrete, plaster, wood finishes, mechanical, and electrical. The model also includes scaffolding. Statistics: Number of 3D components: 210; Number of polygons: 325,000; Number of activities: 667.

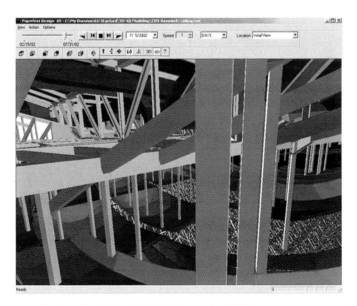

Figure 24. Detailed Hall Ceiling Model. In early 2001, we are constructing a fourth model to help with the detailed planning of the complex concert hall ceiling installation. Statistics: Number of 3D components: 180; Number of polygons: 520,000; Number of activities: to be determined.

CHALLENGES ENCOUNTERED WHILE BUILDING THE 4D MODELS

The construction of the models posed a number of challenges related to the geometry, the schedule, and the linking of the geometry and the schedule. In our experience, such issues are quite common during the development of 4D models, especially when the 3D models are created without knowledge of the needs for 4D modeling and construction planning. Another reason for these issues is that the construction of a 4D model requires significant project scope and schedule information. Some of this information is precisely the information that project participants want to develop or refine through the 4D modeling process, and other information is simply not yet available because of resource or other constraints. A valuable contribution of the 4D modeling process is that the process makes it very clear where complete scope and schedule information exists and where additional thinking is needed.

GEOMETRY ISSUES

Inconsistencies: The 3D models from the architect contained some inconsistencies. For example, an object that was on the plaster layer should have been on the gypsum board layer. Such inconsistencies create extra work during the linking of the schedule and the 3D model because the 4D modeler cannot easily identify, isolate, and show the scope of work for a particular activity in 3D.

Lack of data: The surface model models only what is seen. In the case of a wood wainscot on a plaster wall, FOGA modeled the plaster only where the wood wainscot does not cover it. Even though there is plaster under the wood wainscot, it is not modeled. Hence, in those areas, the surface model does not provide 3D components that can be linked to activities. In addition, for some of the scope of work for steel erection the 3D models were also incomplete. The steel detailers took the wire models from the architect, and produced detailed 3D models from these wires. This process was time-consuming, and at the time of 4D model construction, the detailers had detailed only some of the steel for the main concert hall box. The rest of the steel had to have a 3D representation so that it could be seen during 4D model simulation. We created an algorithm to hang a simple rectangular shape on the wires to make the important information visible without overwhelming the software or the user.

Level of detail: Sometimes there is too little detail in the 3D model. The steel 3D model came back from the steel fabricator all on one layer. However, one might want to split primary and secondary steel into two activities, which would make it necessary to have the primary and secondary steel on two layers. In addition, FOGA modeled just the surfaces for the metal skin. A metal skin system requires backing support and clips, which were not modeled, but need to be installed, and should therefore be reflected in the 4D model.

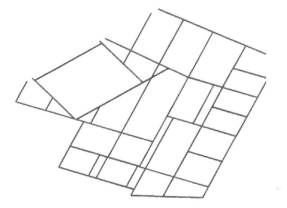

Figure 25.   Simple extrusion on wire.

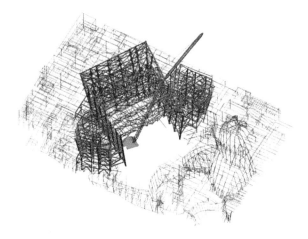

Figure 26.   Mixture of two types of steel models.

Too much data: (Sometimes there can be too much information, which slows down the computational processing of the 3D and 4D models. For example, the steel came back from the fabricator with all the bolts and holes modeled, but we did not need this information for the 4D models the GC wanted to create. Figures 25 to 27 show steel handled at two levels of detail for this project. The resolution of certain situations requires more detail, the resolution of others less detail.

SCHEDULE ISSUES

Inconsistencies: Just as the geometry can be inconsistent with the design intent, the schedule can also contain inconsistencies. For example, the schedule may call for a Concrete Masonary Unit (CMU) wall, whereas the geometry models a cast in

Figure 27.   Steel model from fabricator.

place concrete wall. The inconsistency must be resolved, which, while valuable from the project standpoint, is time-consuming for the 4D modeler.

Lack of data: Some geometry has no corresponding activity. Again, an activity may be required, but resolving this issue requires time and resources of the modeler.

ISSUES WITH LINKING OF 3D MODEL AND SCHEDULE

Inconsistencies: Often, the geometry is defined in ways that conflict with the schedule. For example, the architect defined the geometry by building elements, but the GC places concrete and steel not by element, but rather according to steel sequence. The geometry had to be broken down and recombined a great deal to get a geometrical configuration to match the schedule.

Other data: Cranes, laydown and staging areas, scaffolding, etc. are not part of the architect's design model, but these elements play a large role on the construction site. We had to add these geometries to the 3D model. Figure 17 shows a crane we added to the 3D model to explore the spatial relationship of the crane and its location over time with surrounding work.

Representation of activities with no geometry: Ductwork was not modeled in 3D on most of the project, but the GC was interested to know when and where ductwork was scheduled. A 4D modeler has to be sensitive as to the best way to communicate such activities, by perhaps attaching the activity to a floor slab (as we did), or ceiling framing.

USES OF 4D MODELS

The 4D models supported M.A. Mortenson's four objectives in the following way:

Schedule creation: The GC used the 4D models to assist in planning the lay-down areas for the enclosure contractor, to visualize overall project access at critical junctures in the project, to refine the interior and exterior scaffolding strategy, and to plan the installation of the complex ceiling of the main concert hall.

Schedule analysis: The GC's project management team used 4D models to discover several conflicts in the schedule which were not discovered in the CPM-based Gantt chart. Figures 28 to 30 show snapshops of the 4D models that show particular problems. Figure 28 shows a situation where a CMU wall was scheduled too early while steel was being erected directly overhead. Because the wall that is framed by the steel leans outward the steel erection requires shoring (not modeled), which would not only interfere with the construction of the CMU wall but also cause a dangerous situation. Figure 29 shows an Air Handler Unit (AHU) being installed too late after the steel is completely erected. There would no longer be the access necessary for the large AHU. After consulting with other project team members, the GC decided to leave some of the steel out to make it possible to slide the AHU into the structure at a later date. Figure 30 shows a conflict of scaffolding systems in the same area of the interior hall. The scaffold for the plastering of the walls will need to be removed before the ceiling scaffold can be erected. As a result of the schedule analysis through the 4D model of the interior construction the GC decided to consolidate the scaffolding contracts for the interior hall from three contracts to

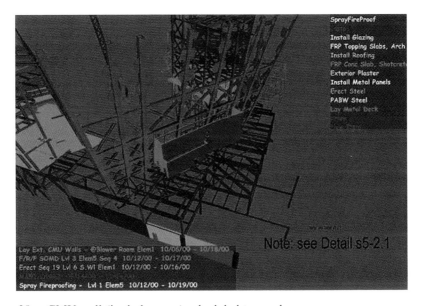

Figure 28.   CMU wall (in dark green) scheduled too early.

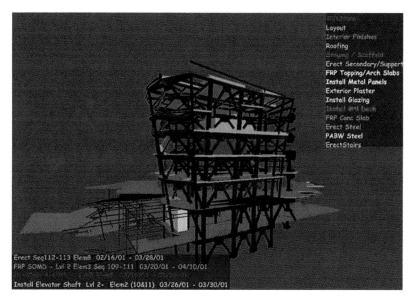

Figure 29.   AHU (shown in red) scheduled too late.

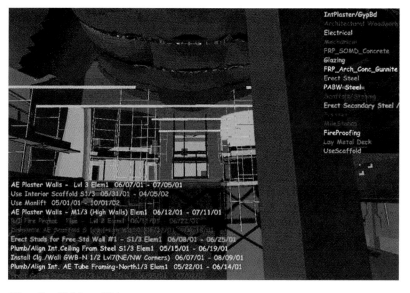

Figure 30.   Scaffolds collide.

one contract. The 4D models supported the discovery of these (and many similar issues) during planning, well before construction started. Note though, that because of the physical and temporal interrelationships between many scopes of work an early detection of potential problems is essential to revise the design or schedule economically. For example, even though the AHU was not scheduled to be installed

Figure 31.   Collaboration in the Virtual Reality Cave.

for many months it was critical to identify potential AHU installation problems prior to work being released for steel fabrication to ensure that the right steel was installed (and not more).

Communication: The GC used the 4D models in training sessions with as many as 40 people, where subcontractors, owners, designers, and the GC reviewed the models and discussed the strategy and constraints for erecting the project. Figure 31 shows a view of subcontractors in a meeting in the WDI CAVE.

Team building: After a 4D review session ended, it was not unusual to have people from different subcontractors remain in the room for an hour or more beyond the scheduled meeting time to discuss issues and solutions to problems or questions identified during the meeting. The GC's project superintendant mentioned that, in a tight labor market, where everyone is committed to too many projects, it is critical to get the attention and collaboration of the subcontractors focused on his project. Given the complexity of the project he wanted to make sure that the subcontractors put their creative energy into improving the construction of his project.

CONCLUSIONS

Our applications of 4D models to construction projects have shown that 4D models help avoid or overcome many of the inefficiencies found on projects today: congestion, out of sequence work, multiple stops and starts, inability to do detailed planning in advance, obstructions due to material stocks, etc. (Koskela, 1999). In all cases except on the Sequus Pharmaceuticals, the Experience Music Project, and the WDCH, the 4D modeling effort required the construction of a separate 3D model because the design had been done in 2D, or the 3D models were not up to date or incompatible with the 4D modeling tools. The schedule information could be used as it was, but often the project team decided to make the activities more detailed to see more detail in the 4D model. As can be seen from Table 1, for many of the benefits the generator of the information necessary for 4D modeling is not the same party realizing the benefits of 4D modeling. Hence, the realization of the benefits of 4D models on projects with a traditional design-bid-build approach often requires extra modeling work. However, the benefits a GC or a subcontractor can realize from 4D models still often outweigh the cost of building the necessary CAD models. On the Sequus project the owner avoided this extra work by awarding a design-build contract to a team consisting of Flad & Associates (architect), Hathaway-Dinwiddie (GC), Rosendin Electric, Paragon Mechanical, and Rountree Plumbing. This maximized the opportunity for each party to enter and maintain the information in the 3D CAD model necessary to realize the benefits. In summary, 4D models allow project stakeholders to work out many design and construction issues in the computer model before actual construction, maximizing project value to owners and making it more likely that the project will be completed as planned and designed.

ACKNOWLEDGMENTS

We are indebted to many professionals and students who have been instrumental in making our 4D modeling efforts successful. We would like to acknowledge the following people in particular: Buddy Cleveland, Jerry King and Kent Simons for the technology support over the years; Jack Ritter, Tom Trainor and George Hurley for getting us started on the San Mateo County Health Center; Todd Zabelle and Greg Silling for sharing construction insights with us over the years; Melody Spradlin and everyone else from the Sequus project for going live; Jim Glymph, Kristin Woehl, and Dennis Sheldon from FOGA for the challenge, fun and excitement of applying 4D models on FOGA projects; Chris Raftery for letting us participate in EMP and Lisa Wickwire for keeping us current with project information on EMP; Ben Schwegler from WDI for his substantial financial and intellectual support and for co-hosting the workshop; and Greg Knutson, Derek Cunz, Jim Yowan, David Mortenson, David Aquilera, and Joe Patterson on the WDCH.

REFERENCES

Akinci, B. & Fischer, M. 1998. Time–space conflict analysis based on 4D production models. In K.C.P. Wang (ed.), *Proceedings of congress on computing in civil engineering: 342–353*. Reston, VA: ASCE.

Atkins, D.C. 1988. Animation/simulation for construction planning. *Engineering, construction, and operations in space: Proceedings of space 88: 670–678*. ASCE.

Cleveland, A.B., Jr. 1989. Real-time animation of construction activities. *Proceedings of construction congress I—Excellence in the constructed project: 238–243*. ASCE.

Collier, E. & Fischer, M. 1995. *Four-dimensional modeling in design and construction.* Technical Report, Nr. 101. Stanford, CA: Center for Integrated Facility Engineering (CIFE).

Collier, E. & Fischer, M. 1996. Visual-based scheduling: 4D modeling on the San Mateo County Health Center. In J. Vancgas & P. Chinowsky (eds), *Proceedings of the 3rd congress on computing in civil engineering: 800–805*. ASCE.

Danhier, B., Massonnet, A. & Verminnen, F. 1994. SG replacement problems anticipated and avoided with 3D CAD. *Journal of Nuclear Engineering International* 39(474): 16–18.

Edwards, R. & Bing Z. 1999. Case study: 4D modeling and simulation for the modernization of Logan International Airport. *Proceedings of the 1997 international conference on airport modeling and simulation: 8–27*. ASCE.

Fischer, M.A. & Aalami, F. 1996. Scheduling with computer-interpretable construction method models. *Journal of Construction Engineering and Management* 122(4): 337–347. ASCE.

Fischer, M., Aalami, F. & Akbas, R. 1998. Formalizing product model transformations: case examples and applications. In I. Smith (ed.), *Artificial intelligence in structural engineering: Information technology for design, collaboration, maintenance, and monitoring, Lecture Notes in Artificial Intelligence,* 1454: 113–132. Springer.

Fischer, M., Liston, K. & Schwegler, B.R. 2001. Interactive 4D project management system. *The 2nd civil engineering conference in the Asian region, Tokyo, 16–18 April, 2001* (accepted for publication).

Fröhlich, B., Fischer, M., Agrawala, M., Beers, A. & Hanrahan, P. 1997. Collaborative production modeling and planning. *Computer Graphics and Applications, IEEE* 17(4): 13–15.

Kahan, E.T. & Madrid, X.H. 1987. Integrated system to support plant operations. *Hydrocarbon processing symposium: 55–60*. ASME.

Koo, B. & Fischer, M. 2000. Feasibility study of 4D CAD in commercial construction. *Journal of Construction Engineering and Management* 126(4): 251–260. ASCE.

Koskela, L. 1999. Management of production in construction: a theoretical view. In I.D. Tommelein & G. Ballard (eds), *Proceedings of the seventh annual conference of the International Group for Lean Construction (IGLC-7): 241–252*.

Retik, A. 1997. Planning and monitoring of construction projects using virtual reality. *Project Management* 3(1): 28–31.

Songer, A.D., Diekmann, J. & Al Rasheed, K. 1998. Impact of 3D visualization on construction planning. In K.C.P. Wang (ed.), *Proceedings of congress on computing in civil engineering: 321–329*. Reston, VA: ASCE.

Schwegler, B., Fischer, M. & Liston K. 2000. *New information technology tools enable productivity improvements. North American Steel Construction Conference, American Institute of Steel Construction (AISC), Las Vegas, 23–26 February: 11-1 to 11-20.*

Staub, S., Fischer, M. & Spradlin, M. 1999. Into the fourth dimension. *Civil Engineering* 69(5): 44–47. ASCE.

Vaugn, F. 1996. 3D and 4D CAD modeling on commercial design-build projects. *Proceedings of computing in civil engineering congress: 390–396*. ASCE.

Williams, M. 1996. Graphical simulation for project planning: 4D-Planner™. *Proceedings of computing in civil engineering congress: 404–409*. ASCE.

BEYOND SPHERELAND: 4D CAD IN CONSTRUCTION COMMUNICATIONS

Dennis Fukai

M.E. Rinker, Sr. School of Building Construction,
University of Florida, Gainesville, FL, USA

Abstract

This study examines the fourth dimension as the product of a fundamental shift in a para-digmatic world view. This shift changes the normal way of "seeing" or visualizing the obviousness of the context of our everyday practices and leads in a different direction with a completely new vision of the processes antiquated by its transformation. This can be seen in the renaissance of closely held ideas that occurred in the change from an oral tra-dition to descriptive diagram, from diagram to written and reproducible text and two-dimensional images, and from simple image to perspective drawings and photographs. These are "visionary" changes that triggered immediate and lasting displacements in our social and technical development.

The shift from three to four dimensions in computer aided design (CAD) does not seem to have had this revolutionary impact even though its value and potential have been made quite clear by a number of researchers. As a consequence, this study explores the context of these new computational tools and how they might be used to enhance the communication process in construction. It suggests that computer mediated communications in construc-tion might be better used to understand the process delineated by a model's construction, where that model is developed as a preview of its construction, "built" according to the same methods and techniques anticipated in the actual project.

Keywords: computer, visual, communications, construction, modeling

INTRODUCTION: "UPWARD, YET NOT NORTHWARD"

In the 1880s, Edwin Abbott wrote a classic book about the idea of dimensions that has been reprinted more than six times (Abbott, 1952). Abbott was a schoolteacher writing about a society of objects that inhabited a land he called Flatland. There

Figure 1.   A circle appears as a line when you live in Flatland.

were no solids in Flatland. Instead, objects existed in two dimensions and were viewed along their edge. This means geometric shapes had lengths and widths, but because they had no height, they always looked like a line. As shown in Figure 1, this is much like a circle first standing on its edge so you can see its three-dimensional face and then laid flat on a table so that only the edge is visible. Unable to rise above this edge view meant circles always appear as a line in Flatland.

This also meant that a square and a circle looked the same. To "see" the difference, one had to feel the edge of an object to know whether it had corners or curves. There was an educated class of citizens in Flatland that could distinguish those shapes without touching them because their eyes had been trained to see the object's outer edge fade slightly into a fog or haze. However, ordinary people did not have the skill to visualize this subtle shift in dimension.

"Space" was therefore flat on a two-dimensional plane and measured in the direction of the length and width of the edge of an object. This meant objects could move north, east, south, and west according to two axis, but it took great skill and a fundamental understanding of the nature of this two-dimensional "space" to move around. Abbott notes there was a "southward" attraction that could be felt in some regions of the plane, but since the edge of an object was always a line, any view of distance had no perspective. In other words, an object could be near or far, polygon or curved, or open or closed, as one moved from place to place. Everything looked the same and it took education and training to read two dimensions and understand the subtle variations of the forms it contained in order to navigate without getting lost.

A sphere called the "stranger" came into Flatland. When it arrived, it introduced another dimension: up and down. From the Flatlander's point of view, a sphere moving up and down through the edge view of the two-dimensional surface of Flatland appeared as a line that grows shorter and shorter until it disappears. This meant that a three-dimensional object like a sphere might look like any other circle or square, but it could actually rise above its two-dimensional plane and disappear as illustrated in Figure 2. No other object in the world view of Flatlanders could shrink, stretch, or disappear in and out of its restricted two-dimensional view like a three-dimensional object.

Even more amazing to the inhabitants of Flatland, was that the very notion of up and down meant that things that were once secret or hidden when viewed from their edge were now exposed when viewed from "above." This meant the inside of a two-dimensional circle could be seen from this other dimension. Anyone able to see in this new dimension could therefore see things that were once considered

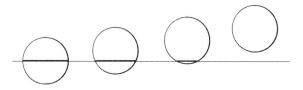

Figure 2.   A sphere appears as a diminishing line when it moves through Flatland.

private and privileged. The idea of a new and elevated perspective changed the fundamental concepts of their world. It required a paradigmatic shift in their way of thinking that made all the "substantial realities" of their world view "appear no better than the offspring of a diseased imagination, or the baseless fabric of a dream" (Abbott, 1952).

Of course, the leaders of Flatland refused to believe in the possibility of another dimension and prohibited all discussion of its existence throughout their land. The idea was simply too disruptive to consider because they were locked into their own restricted, but well-ordered, world view. To see another dimension meant highly skilled citizens had to change their way of seeing. This is not easy for anyone to do, not only because it is disorienting, but also because it calls for a perceptive displacement in a way of living that is not easy to accommodate.

TWO, THREE, AND FOUR DIMENSIONS?

This is the same difficulty many students face when they first look at a complex set of two-dimensional construction drawings (Wilson, 1997; Wei & Gibson, 1998). The idea that a three-dimensional object can be projected onto a collection of two-dimensional planes is contrary to their view of their world. As shown in the example in Figure 3, plan-reading calls for training in order to "see" the shapes and images represented by the lines and symbols laying so narrowly defined on the surface of a piece of paper. Students eventually learn to read plans, but they only learn to see them in three dimensions after they have had a good deal of construction experience.

In practice, the relationship of two-dimensional drawings to three-dimensional space in the design of buildings is less of a challenge. This is because most floor plans require little more than the ability to visualize a vertical extrusion of a collection of lines, certainly not much of a challenge for designers to draw, and even less challenging to visualize before construction. And when spaces are stacked one on top of another in multiple stories, they most often become a series of identical floors, nothing more than a vertical collection of the same extruded two-dimensional spaces. The restrictions of our perceptions as designers and builders therefore seem to confine us to spaces that are relatively simple to draw, visualize,

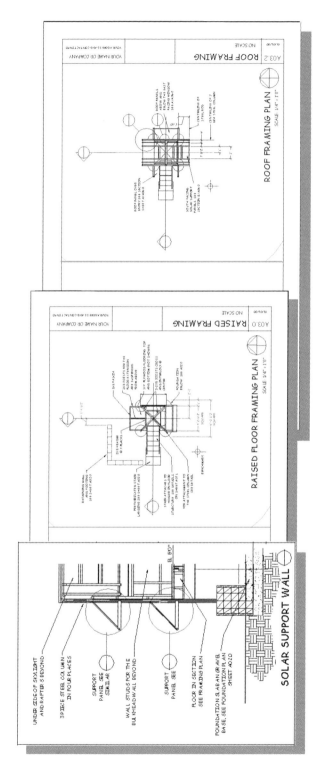

Figure 3.   It takes experience to see in three dimensions.

and build. There are exceptions, but when the imagination of an architect or engineer produces something more complex and three-dimensional than the ordinary, the result is difficult to understand and usually more costly to build. The variation from the norm requires a higher level of interpretation and the result can be at once disturbing and exciting.

This odd perceptive shift from three to two and back to three dimensions also constrains the potential of the buildings that we build. A designer imagines space in three dimensions, but then must translate that space to two and the builder must then take the two-dimensional drawings and transform the lines back to the third dimension.

FOUR DIMENSIONS AND MORE INFORMATION?

This becomes even more disorienting when we introduce the idea of a fourth dimension. Now computers automate much of the information found on a two-dimensional drawing. For some, any computer aided drawing (CAD) introduces a fourth dimension when it is correlated with computer-generated data not normally found in hand drawings (Wright, 1994; Vaugn, 1996; Cardone et al., 1999). However, others point to the obvious flaws in the assumption of this dimensional shift when applied to CAD documents (Shah & Wilson, 1988). If information is a broad category of "stuff" or "chunks," it can in itself describe multiple dimensions of perception (Hofstadter, 1980). In the same way, written descriptions, perhaps associated with two-dimensional diagrams, can bridge multiple dimensions in a way that only a writer can describe. It follows then that a construction document, with the addition of a description or sequential representation that was already inherent in that document, does not necessarily approach a true fourth dimension (Shah & Wilson, 1988; Falcioni, 1999). For example, it is hard to argue that a time sequence image of two-dimensional images shown in Figure 4 represents three or four dimensions.

Similarly, by associating the two-dimensional symbols for chairs, desks, lamps, and credenzas as informational "Blocks" in a CAD program, a furniture layout can be made to automatically sort, count, and list the quantities of the objects that have been inserted into the two-dimensional plan. Is this three or four dimensions?

Of course, the apparently flat furniture plan has a very important third dimension. Walking through the spaces before the installations would show vertical dimensions of counters, furnishings, equipment, steps and stairs, and the height on the walls where signs, clocks, or coat racks are to be placed. There is no doubt that the representation of three-dimensional space and its associated information are well served by the automated references that can be generated by a database, however, it begs the question: do two-dimensional diagrams and symbols represent three-dimensional objects with the automated data defining a third dimension?

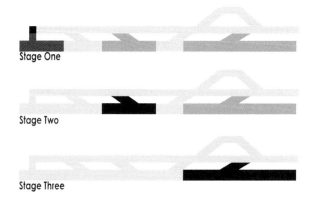

Figure 4.   Three phases for the installation of concrete in a runway construction represent four dimensions.

Most would agree that a two-dimensional diagram of a three-dimensional space is not in itself three-dimensional. A three-dimensional document must be drawn in three dimensions to visually break from a two-dimensional plane (Wei & Gibson, 1998). After all, there are no true vertical relationships in the average extruded floor plan, and furniture installations can most often be completed with no special visualization skills. In practice, the actual construction may not even follow the original plan, primarily because placement will be adjusted according to the way objects "fit" within the assembly. The perceptive reality of the plan will thereby be ignored once its diagrammatic representation is over shadowed by the actual space. In other words, once we see the real thing, everything else is a "baseless fabric of a dream" (Abbott, 1952).

With the introduction of 3D CAD programs, construction models can now be built to meet this same perceptive challenge (LaCourse, 1990). These are models assembled from the three-dimensional pieces of a total structure. As shown in Figure 5, the combination of the assembly of a building as solids and the ability to view these solids from many viewpoints greatly enhances a constructor's under-standing of spatial relationships, construction details, and fabrication techniques.

At the same time, it is difficult to actually construct an object from a three-dimensional model even though, as shown in Figure 6, annotations can be added to explain the construction. This is because details about materials, dimensions, and specifications are still required to build the object on a construction site. To meet this challenge, some innovative practitioners have begun to use three-dimensional models to create two-dimensional drawings (Wilson, 1997). In AutoCAD2000, operators can use "layouts" to convert models to two-dimensional diagrams, annotate and dimension them for use in the actual construction. The perceptive shift to a three-dimensioned construction model therefore reverts by default to the standard two-dimensional drawing.

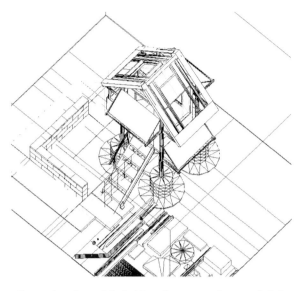

Figure 5.   Three-dimensional models bridge the perceptive gap left by two-dimensional drawings.

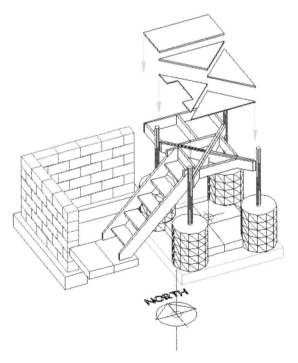

Figure 6.   Three-dimensional models can be annotated, but it is difficult to show the kinds of layout dimensions necessary for actual construction.

Figure 7.   CAD–CAM models used in rapid prototype development embody time in the manufacturing process.

There are exceptions, of course. For example, objects like machined parts can be fabricated directly from the three-dimensional model using CAD–CAM and CNC equipment (Kamarani, 1999). As the result of this direct link between the computer model and the fabrication of a product, an industry of rapid prototypes and integrated design and manufacturing method analysis has emerged. This includes the notion that computer modeling can include a sequential analysis of time as a fourth dimension (Potter, 1998).

The kind of equipment shown in Figure 7 shows how computer models can define both the shape, sizes, and sequences of the manufacturing process, as turns on a lathe, cuts on a milling machine, or holes drilled. This suggests that time is evident in the machine's interpretation of the three-dimensional model. In other words, since the results emerge from the computer model through the mechanisms of a machine, the production "process" is defined by the construction document (Jerrens, 1999). In the same way, the plotted output of a three-dimensional model in a CAD program includes time as part of its documentation, if the process and sequence of the application of ink to paper for the resulting image was specifically defined as the output of the computer model. These time relationships between model and fabrication point to the importance of time in any description of the fourth dimension. Again, the idea raises the question: where is time in this relationship? Does time occur during the actual output as a production process? Or is it embedded in the informational description added to the representation of that process in the drawing or model itself?

When we consider the construction of large objects like buildings and other engineered structures, the representation of time in a computer model becomes even more blurred. First, is an isometric or perspective drawn flat on a piece of paper or visible on a computer screen three-dimensional? And if "chunks" of information are added to the isometric drawings can it be said that this information represents another dimension? It seems logical that the result is simply an extension of the same information that was already resident in the drawing. Why then is the addition of time, in a similar image or visual composition, not simply another informational layer? How does time extend the underlying model into a revolutionary fourth dimension?

TIME AS A DIMENSION

Defining the fourth dimension is a matter of dimensional relativity. Abbott wondered if the fourth dimension had anything to do with an unseen axis in three-dimensional space. He saw a mathematical progression. If a non-dimensional "point" has a single point, a one-dimensional "line" has two points, a two-dimensional "square" has four points, and a three-dimensional "cube" has eight points; then the fourth dimension might be some extension of an object with 16 points. Abbott thought that within the perceptive paradigm of three-dimensional space the fourth dimension might be something he called "extra-solids" produced by "motion of the solids" and "double-extra solids" that result from the "motion of the extra-solids through space."

This analogy of solids in motion through space is interesting when we consider that Abbott was writing these words when Einstein was a child, and space and time relationships remained to be hypothesized and tested as a theory of relatively. It was with Einstein's work that the idea of a three-dimensional world took on the fourth dimension of time. For Einstein, objects no longer simply exist as solids, instead they were part of a continuum of time, perpendicular to the space created by the juxtaposition of the mass of that object. Time is thereby measured by the movement of light and is affected or changed by mass to produce a series of "relativities" defined by the position of the observer.

Non-physicists have taken the notion of time as the fourth dimension and used time in its simplest form to describe a sequence of passing events or phases. In this interpretation, time can be as simple as a series of photographs that capture a particular event. However, this does not seem like an elegant interpretation of the fourth dimension. Are a series of photos showing a sequential event in Figure 8, four-dimensional?

Time can also be shown in a series of model images as the phases of an object's production, operation, deterioration, and/or maintenance (see Fig. 9). In this model, "time" is embedded as a representation of the "motion" of an object in space. The result is that the fourth dimension in CAD has come to be understood as the visual representation of time in the form of images of the phases of the evolution of a three-dimensional model. Others argue we cross the threshold of a

Figure 8.   A sequential series of photographs as a four-dimensional image.

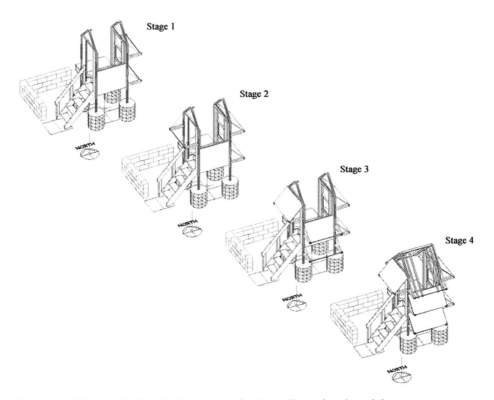

Figure 9.   Time as the fourth dimension of a three-dimensional model.

fifth dimension when layers of information are added to a four-dimensional model (Cardone et al., 1999).

When we think of the complexity of Einstein's theory, the idea that time can be represented in a three-dimensional model as the sequence or phases of a construction, does not seem to reach its full potential. After all, his theory is that the space–time relationship of the fourth dimension is affected by the mass of the solids. In fact, the energy produced by the relationship of mass and time is a derivative of this relationship. This means that the presence of the solid object must distort and constrain time according to the point of view of the observer. Accordingly, the space–time dimension is a continuum that includes the observer; it is the relative position of the observer within this continuum that defines the perceptive shift in that person's view of both time and space.

It could be said that a sequential series of three-dimensional images are nothing more than additional information about the same three-dimensional object, even when the motion associated with the sequence discloses a new perspective (Yamaguchi & Liu, 1998). Time existed or could have existed in the original image as an annotation, it is only its interpretation or visual representation that has been enhanced. This of course includes notes about movement, flow, and events, past or imagined, of the actual object in a photograph or its representation in three-dimensional space.

It seems then that relationship of time and space in 4D CAD may therefore play to the immediate application of sequential modeling without looking to its full potential. There is no doubt that understanding the evolution of an object, either to document its construction or to visualize some aspect of the design, is important, however, this may be a narrow view of a larger space–time relationship.

ONWARD THROUGH THE FOG

Perhaps the fourth dimension is not northward or upward, but onward through the fog of uncertainty. Consider that if the relative position of the observer in a space–time continuum changes the perceptive results of both space and time for that observer, the true fourth dimension may depend on how the observer visualizes changes in the model over time. This would make change important in understanding the continuum of these dimensional boundaries.

This is evident in the work of a number of researchers. For example, the Center for Integrated Facility Engineering (CIFE) used a construction method model (CMM) and expanded it into something called "Collaborative 4D CAD" (Aalami & Fischer, 1998). A CMM is a variation on a phased model that is used to analyze the construction process through visualization. It asserts the inclusion of a fourth dimension because its CMM analysis focuses on a process that relates planning and management decisions directly to the sequence of the building's construction.

much the same way they might be found in the field. In fact, many of the changes made to the model were visible in the completed building.

These changes and the conflicts that were discovered in the pre-construction of this building were valuable as a precursor of the actual construction process, but the pattern of communications about the construction of this model emerged as the lesson learned from this experiment. The communications log showed that time and space were in fact changed by the construction. In the analogy of Einstein's theory: time, or the warp of time, occurred as the "mass" of the construction model disturbed or changed the vector of assumptions made from a particular point of view. In other words, as one might suspect, a series of errors, omissions, and requests for information indicated the need for change orders that adjusted both scope and schedule.

Though a simple experiment, and certainly not intended to be the kind of proof that Einstein sought for his hypothesis, it seems like this idea might in fact point in a different direction for what is more commonly thought to be 4D CAD. It also suggests a more a robust use of computer modeling, giving a "hands-on" experience from which to learn about constructing buildings while simultaneously previewing the quality of the construction documents. What is important is that the resultant model was 4D CAD, but unlike a sequential visual explanation like the series of images shown in Figure 14, time became the distortions and changes that were visible in the pre-construction communications.

First, the modeling effort turned up errors and omissions in the construction drawings. This would be important to limiting requests for information, clarifications, and change orders that would have occurred during the actual construction. Second, the model helped to understand the quantities and methodologies associated with the materials and labor that would be part of the same process. In practice, this would help verify estimates and strengthen confidence in work plan simply because the construction process could be evaluated from the construction model rather than traditional two-dimensional drawings and specifications. Third, the resultant model provided an archive of graphical images indexed according to the pieces of the construction that would support future construction communications. This includes zoomed close-ups, three-dimensional details, time sequence or phased images, and two-dimensional projections used as shop drawings or field clarifications in both the building and the falsework, formwork, and special structures for the project.

The pedagogical opportunities are equally exciting. One of the conclusions of the students involved in the experiment was that the experience had taught them a lot about constructing a building. They pointed out that they were also able to contextualize many of the things they had learned in other construction classes. For example, as shown in Figure 15, the construction of the running bond and reinforcing on a CMU wall is quickly modeled in a class assignment.

If this is true, a similar 4D CAD experience for a specially designed building might be useful as a tool in construction education. The overall learning strategy

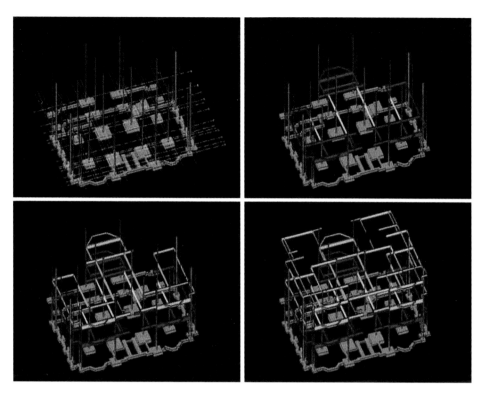

Figure 14.   Actual construction is enhanced by communications during the pre-construction modeling process.

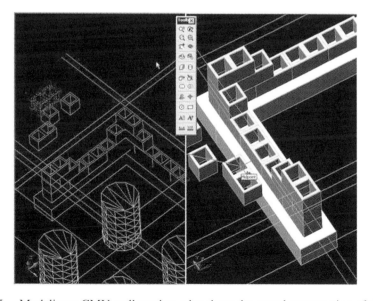

Figure 15.   Modeling a CMU wall teaches a lot about the actual construction of that wall.

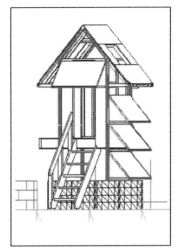

Figure 16.   Construction model and physical construction can be used in combination to test ideas and planning.

of such a tool might include a construction project that allowed students to work in teams to complete the assembly of a virtual project. Students would then break the work down and interact with each other to build the virtual model within the time constraints. In these interactions, students would have to read the two-dimensional plans, use hand-drawings to communicate their ideas in face-to-face meetings, build portions of the building, and transfer their ideas to other team members as email attachments. This could be supplemented by a full size version of the construction model to give students a hands-on feel for the actual construction (see Fig. 16).

After completing the virtual construction they strengthened their plan-reading skills and their ability to visualize a three-dimensional object from two-dimensional plans. This suggests they could use hand-drawings and computer images to spontaneously visualize the construction. And perhaps most importantly they would be able to review the entire process to understand the context of their actions and how a similar pre-construction effort might be used for other projects.

PRE-CONSTRUCTION AS PRE-COMMUNICATIONS

In an informal market study, the notion that buildings can be "pre-constructed" on a computer from a set of construction documents using the same labor and materials that would be used on an actual job site seemed to stir the interest of construction managers (Fukai, 1996a). In its simplest form this means prethinking a building prior to its construction; but it also means using computer modeling to improve job safety, simulate staging for difficult operations, coordinate complex lifts, discover errors and omissions in a set of drawings early in a project, analyze change orders, value engineering, and improve client communications in standard construction reports.

It is important to emphasize that a pre-construction model is not the same as an architectural or engineering model. Architectural models help designers, builders, and owners visualize the spatial or aesthetic relationships in a building and are often used to illustrate phases or alternate finishes. Similarly, engineering models are programmed to illustrate the reaction of the structure to dynamically loaded forces placed on its frame or other structural elements. By contrast, a pre-construction model is an anatomically correct representation of all the pieces of the construction. This includes both the building and the falsework and formwork that will be required for its construction. Pre-construction models require a complete understanding of the construction process, including the methodology and techniques that will be unique to any particular project. In other words, the model's assembly must include all the details of the construction and follow the same methods used in the actual production process.

Thus the model becomes an instrument of communications, rather than visualization. To achieve this potential, it must be built in a way that allows clients to archive and manipulate images that will be useful during the project's actual construction. This means the resulting model is not as important as the communications process that occurred during its virtual construction.

To capture this process, the modeling effort should therefore be divided into subcontracts that parallel the work breakdown structure for the actual building. This delineates the responsibility for the pre-construction and sets up the context for communications. Using a common workpoint to coordinate the construction in much the same way it would be done on a job site, each piece of the building could therefore be identified according to its subcontract and placed on a distinctly controllable layer. This also allows rapid development among multiple team members and the ability to deconstruct and analyze the building piece by piece.

Once the model is complete, it then becomes important to extract the data that it contains. One way to do this is to use something called a data-theater shown in Figure 17. The data-theater was introduced as a concept at the ACADIA (Association of Collegiate Schools of Architecture) convention (Fukai, 1996b). It is basically a computational "black box" that surrounds the model and acts as a graphical interface to a software engine. Clicking on the box initiates macros that deconstruct the model according to preset planes that surround and slice through its three-dimensional form. Images and diagrams can then be extracted from the model by a software engine to facilitate the actual construction project. This "engine" is a simple set of customized macros that does not involve high level programming skills. This is important because it will have to be "tuned" to each particular project.

Another method is to use the model as a hypergraphic interface (Fukai, 1996b). Clicking on pieces represented in the interface zooms in on detailed "layers" of

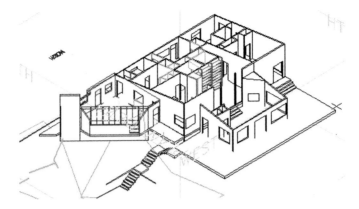

Figure 17.   A data-theater surrounds the model to annotate layout planes.

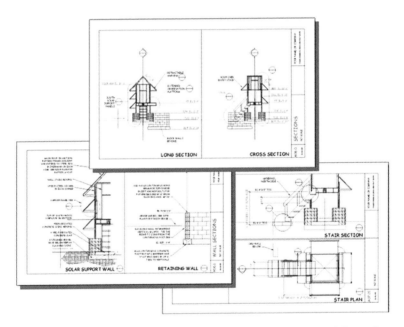

Figure 18.   Two-dimensional construction drawings can be generated from the four-dimensional modeling process.

supporting visual information. For example, excavation, backfill, concrete, and reinforcing steel can again be represented in the model as subcontracts on different layers so that they can be displayed separately or in different combinations. The organization of these layers and the way they are "called" by the graphical links would allow users to dynamically deconstruct the model by moving toward ever more detailed representations of the construction.

Both of these methods point to a graphical interface that leads to an infinite collection of dynamically generated "snapshots" of the pre-construction model.

This means that there is no part of the building, before or after its construction, that will not have a graphical representation available for inclusion in daily or monthly reports, presentations to clients or subcontractors, websites, service manuals, and ongoing management and repair of the facility. In addition, many parts of the building will have detailed annotated two and three-dimensional layouts that can be used to direct the construction. These layouts are generated from the three-dimensional drawings (see Fig. 18).

CONCLUSION

The challenge is to explore the modeling process in the context of an actual construction project in order to search for the revolutionary insight that must be part of what is truly the fourth dimension. Missing from a three-dimensional model is the communications that is embedded in the construction simulation.

The perceptive shift is not in our view of the model, but in a sideways view at the communications associated with that model's construction within its virtual environment. Extracting the graphical data associated with the flow of that communication provides the constructor with an archive of visual explanations that could thereby anticipate the interactions that will occur in the actual construction process. In this way, pre-construction becomes pre-communications, and the effort suggests a new graphical "intelligence" that might be added to support an industry of complex practices.

In conclusion, if there is to be a radical shift in our world view as we move from three to four dimensions, perhaps it must occur in the kind of communications represented by the pre-construction modeling process rather than the model itself.

REFERENCES

Aalami, F. & Fischer, M. 1998. Construction method models: the glue between design and construction. *Proceedings of the 1998 international computing congress on computing in civil engineering: 376–378*. Boston: ASCE.

Abbott, E.A. 1952. *Flatland*, 6th edition. Dover Publications.

Cardone, F., Francaviglia, M.& Mignani, R. 1999. Five dimensional relativity with energy as the extra dimension. *General Relativity and Gravitation* 31(7): 1049.

Falcioni, J.G. 1999. Managing product life cycles. *Mechanical Engineering*, CIME, 121: 4 (editorial).

Fukai, D. 1996a. A WORLD of data: an animated hypergraphic construction information system. *Presentation to the Association for Computer Aided Design in Architecture, Tucson, AZ, October 1996*.

Fukai, D. 1996b. Real-world, real-time, real-fast: using a trainer in a computer mediated classroom. *A presentation as a fellow to the Materials and Technology Institute of the Association of Collegiate Schools of Architecture, Berkeley, CA, 1996*.

Fukai, D. (unpublished paper). *Current Research: Insitebuilders.com.* M.E. Rinker School of Building Construction, College of Architecture, University of Florida.

Hofstadter, D.R. 1980. *Godel, Escher, Bach: An Eternal Golden Braid.* Vintage Books.

Jerrens, K.K. 1999. NIST's support of rapid prototyping standards. *IEEE Spectrum* 36(2): 38

Johnson, D. 1999. Discuss and change models in real time. *Design News* May 3, 1999: 96–101.

Kamarani, A.K. 1999. *Direct engineering: toward intelligent manufacturing.* Klumer Academic, Monograph.

LaCourse, D. 1990. How solid modeling previews the future. *Design News* 46(10): May 10, 1990: 90–92.

McKinney, K., Kim, J., Fischer, M. & Howard, C. 1999. Interactive 4D-CAD. *Computing in Civil Engineering*: 383–389.

Potter, C.D. 1998. Process control CAD/CAM's newest tool aims to oversee both the data and the methods used to design complex assemblies. *Computer Graphics World* 22(8): 69.

Shah, J.J. & Wilson, P.R. 1988. Analysis of knowledge abstraction, representation and interaction requirements for computer aided engineering. *Computers in Engineering, 1988 – Proceedings.*

Vaugn, F. 1996. three-dimensional and 4D CAD modeling on commercial design-build projects. *Computer in Civil Engineering* 1996: 390–396.

Wei, D. & Gibson, K. 1998. *Computer visualization: an integrated approach for interior design and architecture.* McGraw-Hill.

Wilson, J. 1997. *AutoCAD: a visual approach.* Autodisk Press.

Wright, V.E. 1994. 4D CAD. *Heating, Piping, Air Conditioning* 56 (July 1984): 41–53.

Yamaguchi, T. & Liu, H. 1998. Computational visualization of external and internal biological flows with fluid-wall interactions. *Advances in Bioengineering* 39: 127–128. American Society of Mechanical Engineers: Bioengineering Division.

FULLY INTEGRATED AND AUTOMATED PROJECT PROCESS (FIAPP) FOR THE PROJECT MANAGER AND EXECUTIVE

F.H. (Bud) Griffis, Carrie S. Sturts

Department of Civil Engineering and Engineering Mech,
Columbia University, NY, USA

Abstract

Fully integrated and automated project process (FIAPP) is an acronym suggested by the research committee of the Construction Industry Institute. A schedule linked to a three-dimensional model (or 4D CAD) is a component of a larger FIAPP picture. This paper briefly describes 4D CAD in the context of the larger picture of FIAPP and the three-dimensional computer model. The focus of our research has been on the use of three-dimensional computer models for construction management. Furthermore, we focus on the industrial process or commercial power projects because they are routinely designed in three dimensions. This paper introduces some of the benefits of using three-dimensional models for construction and the conclusions developed in a three-year research project into FIAPP and 3D CAD that relate to 4D CAD development and usage. This discussion is illustrated with a case study project: the construction of an Air Separation Plant in Baytown, Texas. Finally, this paper discusses a possible course outline to teach project managers and project executives how to benefit from using FIAPP in the management of the construction project process. Unless project personnel actually have hands on use of the model, it loses its value as design is turned over to construction. Therefore the state of the art can only be advanced if project personnel feel comfortable operating in a FIAPP environment.

Keywords: FIAPP, training, advantages, 3D CAD

BACKGROUND

The process and power industries routinely use 3D CAD in design and construction; however, most building and heavy construction projects are still being designed and constructed using two dimensions. Given that contractors work from two-dimensional drawings, unless the designers are proficient in three-dimensional design, there are probably few benefits to be gained from three dimensions for relatively simple structures. The benefits of three-dimensional design in residential and commercial buildings have not been shown. On the other hand, the industrial process and commercial power sector of the architectural/engineering/construction industry routinely use three-dimensional computer models for the design and construction of plants and facilities, and the benefits have been well documented (Griffis et al., 1995). Current research efforts are being made to integrate all aspects of the project process (preplanning, design, construction and start-up) using three-dimensional computer models and integrated databases. This paper will discuss 4D CAD and its role in the integration process and the functions of the integrated system in the construction process.

WHAT ARE THE BENEFITS OF USING THREE-DIMENSIONAL MODELS ON THE CONSTRUCTION SITE?

The authors were involved with the Construction Industry Institute's (CII) research in the use of three-dimensional computer models for construction management applications spanning from 1993 to 1995. The study consisted of three parts. First the researchers used questionnaires to investigate the perceived benefits and impediments to using three-dimensional models in the management of construction. Second, they performed statistical studies on 93 projects that used three-dimensional models in the management of construction to varying degrees. Finally, the research team used a case study project to judge the reality of the statistics results. Some of the results are as follows:

Most common usage
- Checking clearances and access
- Visualizing details from non-standard viewpoints
- Using model as reference during project meetings
- Performing constructability reviews

Greatest perceived impediments to the use of 3D in construction
- Undetermined economic impacts
- Inertia
- Lack of trained people
- Cost was perceived as an impediment only by non-users

Perceived benefits by users
- Reducing interference problems
- Assisting in visualization
- Reducing rework
- Improving engineering accuracy
- Improving jobsite communication

Differences between only 2D and "average" to "very good" use of 3D*
- 5% reduction in cost growth
- 4% reduction in schedule slip
- 65% reduction in total rework

---

* Benefits were quantified by the statistical study (Griffis, 1988).

These benefits are impediments that may specifically apply to 4D CAD in varying magnitude; however further research should be conducted to isolate the benefits of 4D CAD for various project types and at different management levels. The case study project was used to actually perform cost estimates of the benefits as they occurred in the field. Those direct cost benefits exceeded those predicted by the statistical computer models. (Griffis et al., 1995).

WHAT IS FIAPP?

We have come to the conclusion that the greatest benefits from the three-dimensional computer model come from the integrated databases and not *just* from the three-dimensional computer model. Four-dimensional computer models are part of a larger picture. Researchers at CII have coined the term fully integrated and automated project process (FIAPP) to describe this bigger picture. FIAPP is not a software system per se, but an idea about the future computer data systems that will support a project from inception to start-up and beyond. FIAPP describes how information will flow automatically from one system to another, from one project participant to another, from owner to designer to fabricator to constructor. FIAPP and the relationship of the three-dimensional model to the other systems is still being debated.

When we use the term FIAPP, we include all activities in the pre-project planning, the design, the procurement, the construction management, the start-up, and the operations and maintenance phases.

In the ideal FIAPP (Fig. 1), all systems are integrated from payroll to job costing to scheduling to the design systems. Of most interest to construction managers

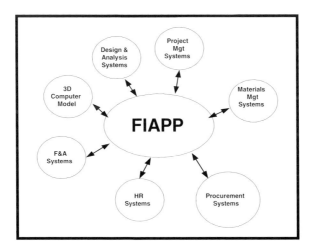

Figure 1.   The fully integrated and automated project process.

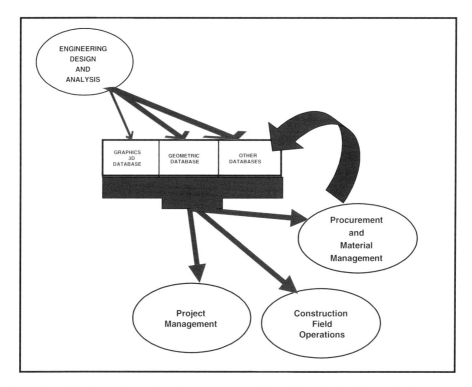

Figure 2.   The three-dimensional computer model and its relationships with the design and construction process.

is a subsystem of FIAPP consisting of the three-dimensional model, the PM systems, the procurement systems and the material management systems. It is within these systems that we will consider 4D CAD. The three-dimensional computer model (Fig. 2) progresses from the design phase with output to the project management systems, construction field operations and the procurement and material management systems. These systems in turn provide feedback to the model and/or its associated databases.

Many feel that the three-dimensional computer model is but one of the databases associated with FIAPP. Others feel it is the hub of the system through which the other databases are accessed. Figure 3 illustrates these two different integration schemes.

For many projects, much of the design and procurement is initiated long before a detailed three-dimensional computer model is available. The design often starts with the process flow diagrams (PFDs) and the piping and instrumentation diagrams (P&IDs). The front end engineering design (FEED) generally consists of PFDs, preliminary P&IDs, multiple simulation cases and cost comparisons, detailed equipment data sheets, request for quotation responses, general equipment and piping layouts and early three-dimensional computer model reviews

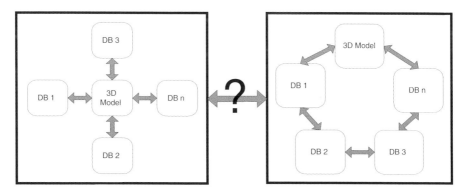

Figure 3.   Spoke concept versus the wheel concept of FIAPP.

Figure 4.   Piping planning.

(Carell, 1999). Most of the piping, piping accessories, and major equipment items are developed in this stage and some of the procurement may be initiated (Fig. 4). This initial development supports the wheel model for FIAPP; however, there are those who feel that once an initial model is developed, it should be the center and organization format for the project data. FIAPP is still in development.

4D CAD AND FIAPP

The following rendering is of the case study project (Fig. 5) upon which this paper is based. 4D CAD was not used in this case study project although it was tried.

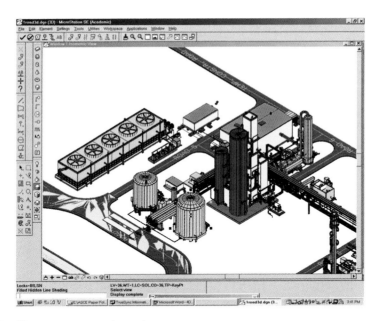

Figure 5.   Baytown case study project.

There were numerous reasons that it was not used. This project was the project team's first attempt to use three-dimensional computer models in the management of construction. The way in which the three-dimensional model was developed greatly inhibited the attempt to use 4D CAD. We found that 4D CAD was impossible to use with this model. The system was designed in Intergraph PDS and third party add-ins. The organization of the model was done so as to preclude separating the commodities and groupings into activities. First, the underground civil work was not modeled. (We have found this to be a shortcoming in most models and will recommend that civil work be modeled in the future.) The civil work could not be scheduled, however, from the model. Secondly, there is the issue of the plant design area relationship to the construction area. Plant design areas take form during front-end project planning. The areas are developed using associated equipment usually by the process section or by the process system. If areas are system based, they can overlap. Construction areas should use the same geographical boundaries as the plant design areas. Construction zones are used to identify physically defined subsets of design areas and are identified by design element attributes. Specific design element attributes can be used to identify the specific subcontractor work packages. Work packages are usually based on geographical boundaries and specify subcontractor data and system tie-in packages. Work is usually controlled by line list data and subcontract data by specific spool number. A back-to-front schedule establishes the design release by construction area or sub-area and is issued as a design area package (Hall, 1999). Finally, the project team did not feel that the value of using 4D CAD on the project site would justify the effort of trying to make the model compatible with the 4D CAD software.

The contractors required two-dimensional paper drawings. The work packages were put together from these drawings and scheduled by experienced engineers. The benefits of the 4D CAD could not be imagined. Based on this fact, we tried to anticipate what the benefits of using 4D CAD on a project this size could be if the model was developed in such a method as to facilitate its use. The following potential benefits were noted:

- If the model were developed early enough, a 4D CAD approach would have been of benefit to permit the Board of Directors to make financial decisions regarding the plant.
- A 4D CAD approach could have been valuable tool for the marketing department in their effort to sell the product.
- A 4D CAD analysis could have helped with the interface with the Exxon plant to which liquid oxygen was to be piped.
- A 4D CAD modeling approach might have found some scheduling errors in the construction of the plant. However, there were no important scheduling delays encountered. Numerous conflicts were found and resolved in the field.

COURSE OUTLINE: USING FIAPP FOR PROJECT MANAGERS AND EXECUTIVES

This course will acquaint the project manager or project executive with the use of three-dimensional computer models to enable an understanding of the evolution of the fully integrated and automated project process. The course will involve computer software. However, it will not be a programming course. It will require understanding of the software characteristics and the manipulation of the software. In this course, we will use mostly software marketed by Bentley Systems. This software is not the only available software that will do the job that we require and it may not be the absolute best software for what we are after. Nevertheless, it is easy to learn and robust enough to serve our purposes. The course is a one-semester course consisting of 45 contact hours. It uses a relatively simple (but not trivial) construction project as a semester project.

Objectives of the course
- To acquaint the project management professional or the project executive with the concept of FIAPP.

NEW CONSTRUCTION MANAGEMENT PRACTICE BASED ON THE VIRTUAL REALITY TECHNOLOGY

Jarkko Leinonen[1], Kalle Kähkönen[1], Tero Hemiö[2], Arkady Retik[3],
Andrew Layden[4]

[1] *VTT Building and Transport, Finland*
[2] *Eurostepsys Ltd., Finland*
[3] *School of the Built and Natural Environment, Glasgow Caledonian University, Glasgow, UK*
[4] *Department of Civil Engineering, Division of Construction Management, University of Strathclyde, Glasgow, UK*

Abstract

This paper focuses on experiences of implementing 4D applications (3D building geometric data + time) to meet the needs of construction companies. The authors have been developing and experimenting with 4D applications based on virtual reality (VR) technology and its integration with other state of the art software. The objective of the paper is to provide understanding for balancing possibilities and challenges of this approach for construction planning and management.

Several case studies with YIT Corporation, a Finnish construction company, have provided basis for many findings to be presented in the paper. This long-term co-operation has covered product modeling, software architecture design, web technology and, more recently, 4D together with VR. First the present problems originating from the current construction planning practice are discussed. This is followed by presenting the possibilities of construction management practice with the aid of 4D—what is needed and what can be achieved. The 4D approach has been demonstrated in the live building construction project, THK office building. The case study covers cost-benefit analysis of applying 4D for construction planning.

Keywords: 4D, virtual reality, construction management, information technology, construction company

INTRODUCTION

Building construction is about transforming the vision and views of a client into a building where the client's needs and objectives are met. This process comprises

75

many stages and phases, iterative activities, and can have large numbers of participants involved having massive information transfer and communication needs.
Furthermore, the practical set up of the building construction processes can vary
greatly from one project to another. A major reason for this is the technical differences between projects. Another main reason is the varying building construction
skills, experience and knowledge of participants. This experience, skills and
knowledge can vary greatly between different organizations involved and require
special attention.

It seems that in practice too many building construction processes are far from
the most suitable processes. This results in clear and easily identifiable high costs,
low productivity growth and poor quality, which, unfortunately, are internationally
common features in building construction. Recent results from several studies show
that deficient management and organization are main causes of these shortcomings
(Koskela, 2000). Additionally, Koskela shows that the main sources for production
defect costs are the design and engineering phases and the production management.
Likewise, deficient production planning has proved to be one of the underlying
reasons for the problems in the production management (Josephson & Hammarlund,
1996). Consequently, it often happens that the design phase in a building construction
project is realized without sufficient constructability consideration. Also, it has been
discovered that the methods for decision support are ad hoc and unsystematic. This
is a root cause for many problems and disappointments in building construction
projects (Laitinen, 1998).

The common feature of the important findings presented above is that they all are
related to building construction management and associated decision-making. In
modern building construction, where power and responsibilities are shared between
key partners, and, where many other stakeholders are involved, the operations management is a real challenge. The overall process from the early project start to hand
over needs to be under improved control at all stages in order to maximize added
value to client. This is principally a task for the construction management practice.

Due to historical preoccupations regarding building construction planning and
management, these processes have been very much *project-realization-oriented.*
This means that practitioners tend to get involved in detailed planning and its
decision-making in a too straightforward manner. However, in practice one can
often encounter unclear or conflicting objectives, high levels of uncertainty relating
to most estimates, communication problems between individuals, unrealistic opinions and a lack of creativeness, flexibility and consensus between various parties.
New methods are needed particularly for reaching an improved level for building
construction management and relating decision-making in an environment where
numerous participants are working together (Turner, 1993; Barkley & Saylor, 1994).
One of these new methods is the use of virtual reality (VR) technology for production
planning and construction management.

One of the benefits that can be gained from VR technology is the possibility for
the 3D visualization of construction plans (Goldstein, 1995). These visualizations can

enable an improved communication over the product and its construction processes (Ogata et al., 1998). It would be much easier to get all key individuals from partnering organizations involved in an improved way and, in particular, to take advantage of their experience and knowledge. The potential problems relating to the efficiency of the current design solution and construction plan can be more easily identified. More detailed benefits have been discussed by Alshawi (1996).

VTT Building Technology has, together with YIT Corporation, been developing and implementing applications based on VR technology and 4D (3D + time) to improve construction management practice. Examples of the areas this long-term operation has covered are product modeling, software architecture design, web technology and 4D together with VR. The results from this co-operation form the basis for this paper. The paper presents different approaches for the implementation of the 4D model. In addition, it is demonstrated how 4D models based on the use of VR technology are created:

- Combining together product modeling technology, scheduling data and VR technology.
- Having a specific building component library which is used step-by-step to build up a 4D model from drawings and scheduling data.
- Additionally, the potential benefits of 4D models are analyzed using a live construction project as a test bench.

3D PRODUCT MODEL AS A STARTING POINT

A building product model is an information base of a specific building (Björk, 1995). These product models can be used by different computer systems to create, edit, store, retrieve and check data about buildings. A leading principle, and promise, is to store all data relating to a specific building into its building product model. As a result the product model would be a common source for all project participants throughout the building life cycle to access building data (Fig. 1).

Development of building product models is strongly facilitated by modern data modeling techniques and tools. A working building product model requires a *product data model*, which structures the information needed to describe a building. In a way, the product data model sets up a standard that forms a basis to integrate together various software tools used for different purposes, in different phases, and by different participants, during the building project life cycle. The building product model captures a lot of different kind of data from which the location and dimension data of each component are only a small portion. However, this data enables the graphical viewing of the resultant product model, or its parts, providing a significant means for communication and teamwork. The resultant product model

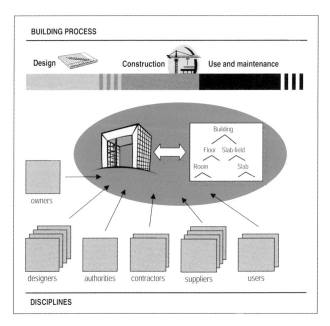

Figure 1.　Product model as a main source for accessing building data by various project participants.

works as a common platform for storing and accessing data for different purposes. Typical examples of various needs for using the data captured in the product model are quantity take-offs, cost analyses, scheduling, resource usage planning and procurement planning.

Another important development principle is the minimization of data redundancy, i.e. the information is stored only once and the documentation and reports are produced from the product model when required. In this way the building product model captures the geometrical representation of the building items, e.g. superstructure, slabs, rooms, walls and doors. This data together with project scheduling data form the starting point for 4D applications (Fig. 2). The latest versions of Industry Foundation Classes (IFC) standardize the data structure of the geometrical representation of the building together with scheduling data.

Product modeling technology can contribute strongly towards advanced partnering and co-operation in construction. Thus, in YIT Corporation, the product modeling technique is being used as a core technology for providing the basic enabling solution for supporting, forming and managing temporary networks of companies and their resources working in a building construction project. The application of this technology is the CoVe model builder (Laitinen, 1998). The CoVe model builder enables the use of heterogeneous data from various project partners for setting up a product model. The model builder CoVe was first developed for the modeling of residential buildings (Fig. 3).

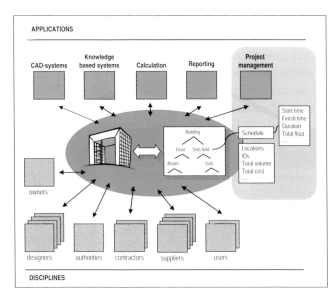

Figure 2.   Using building product model as a means to integrate together various software and contributions by different partners.

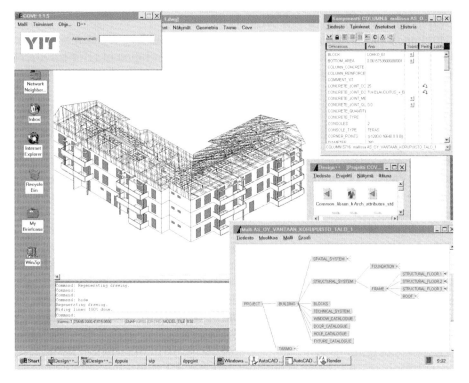

Figure 3.   CoVe model builder: interactive views over product model hierarchy and the resultant building under design activity.

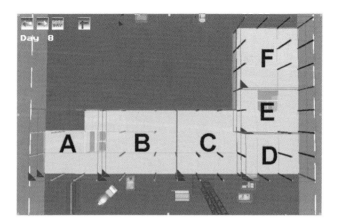

Figure 10.   Division of stories to six sections (a view from the top).

renders the entire building invisible. The program then controls what groups are visible on a specified day. The building is constructed on a story-by-story basis in the technological sequence. Columns and the concrete staircases are erected first, followed by beams and then by hollow slabs. The façade elements are assembled last. This activity is performed in the bottom–top direction.

DEVELOPING THE MODEL

Dividing the building

The first six stories of the building are identical, so they are divided in an identical manner. Each floor is divided into six sections. This was determined by the calculation of the daily rate of construction for the building. The arrangement is shown below (see Fig. 10). The top three stories each have a different layout. The rate of construction is different for each story. This means that they have a different number of groups, but the principle remains the same.

CREATING THE ELEMENT GROUPS

Not all elements are placed on the same day at a particular part of the building. This means that each type of element must be contained in a separate group. Figure 11 shows the first section of the ground story—columns, beams, staircase and hollow slabs. Note that the columns extend for three stories. Each type of element is contained in its own group, with the exception of the staircase, which is a single entity.

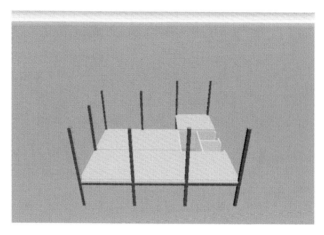

Figure 11. Section 1A of the building.

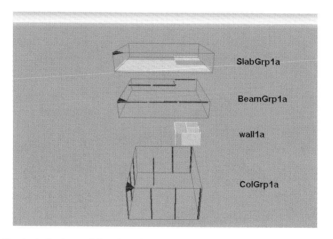

Figure 12. Exploded view of Section 1A.

An exploded view of the same section is shown in Figure 12. The numbering convention is also shown. The group is split into three properties: the type of element, the story, and the section of the story. The area shown is on the first story, and in Section "A".

Note that the group for the hollow slabs sits on top of the other groups. However, it will occupy the same space as the groups for the first section of the second story. Each section of every story shares this arrangement of groups.

In general, the groups for the façade elements do not occupy the same space as the other groups. They are normally placed along the sides of the other groups. However, this is not the case for the first section of the building (Section "A"). This is because that area of the building has an irregular shape. It is best to ensure that different groups do not intersect one another, so a special grouping arrangement was created. The top three stories also contain some unique grouping arrangements.

Creating the elements

The simulation shows only a few of the different types of elements found in buildings. For the sake of graphical simplicity, the majority of elements shown are merely simple cuboids: beams, columns, hollow slabs and façade elements. These elements make up the bulk of the building. They can be modeled accurately, but the simulation would run much slower due to the increased amount of facets on screen. The dimensions of the elements were taken from the plans. Some liberties, however, were taken to increase the modularity of the building, and hence decrease the complexity of it.

The first elements to be placed in the world were the columns. The first column to be placed formed the origin of the building. This is at the bottom, left corner of Figure 10. The beams were simply placed between the columns. The hollow slabs were given the same thickness of 200 mm. These elements were initially modeled as individual slabs, with the dimensions taken from the plans. Yet, it soon became apparent that the simulation would run much faster if the slabs were simplified. Thus, they were changed into long strips of slabs. These could not be allowed to share the same space as the columns, so shorter strips were used to fill in the gaps.

The façade elements were also simplified for the simulation. They are not made of individual panels, as indicated by the plans. Larger panels replace many small ones. The thickness of the panels was set at 300 mm. The concrete staircases started off as four cuboids grouped together as a square. A group enclosed these objects, and they formed a wall group. This was recently changed to a dedicated shape, created with the shape editor. This was done to decrease the number of objects in the world, and hence speed up the simulation.

Once the first story was completed, it was then taken apart. This was in order to create the groups required for each element. Each type of element was temporarily displaced by a set distance from the building. A group was placed around them, and then the group containing the elements was placed back into the correct position. Figure 13 shows how the columns were temporarily moved, so that their groups could be created. The other element groups have already been created and moved back to the correct location in the world.

The groups are then checked for any improper child/parent adoptions (a hierarchy should be defined between different objects). Sometimes, when a group is moved back to its correct position, it adopts one of the other groups. The groups must be completely orphaned. An element group must only contain the appropriate type of element.

Once the first story was checked, it was simply cloned and the next story was placed above the current one. The groups in the new story were renamed to follow the numbering convention previously explained. The first six stories were created

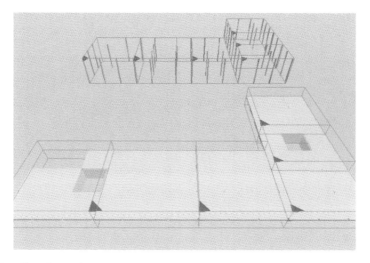

Figure 13.   Creating column groups.

this way. The top three stories were also created in a similar way, but with different numbers and sizes of groups.

THE SIMULATION PROGRAM

The code for the simulation can be classified into three different types: the *main code*, the *icon code* and *background code*. They are all interconnected, by passing counter values to each other.

The *main code* controls the simulation of the building. It decides what sections of the building to display on a given day. The main code also decides what icons and instruments are enabled on the screen display. The code is attached to the "Anchor" object. This is disguised as a portacabin in the VR world (from viewpoint 4, it is visible in the bottom right corner of the screen). The main code is by far the largest code in the simulation.

The *icon code* is used to pass control data to the main program. It is also used to control some of the vehicles in the world, both directly and indirectly. Many of the icons are unavailable at certain times during the simulation. Their availability is controlled by the main code. The icons process the SCL User Functions.

The *background code* is attached to various objects in the world. Some of the objects, like parts of the tower crane, communicate with the icon code and the main code. The cars that travel on the roads around the world have their own predetermined paths. These are controlled by their own code. They have no connection to the icon code, or the main code.

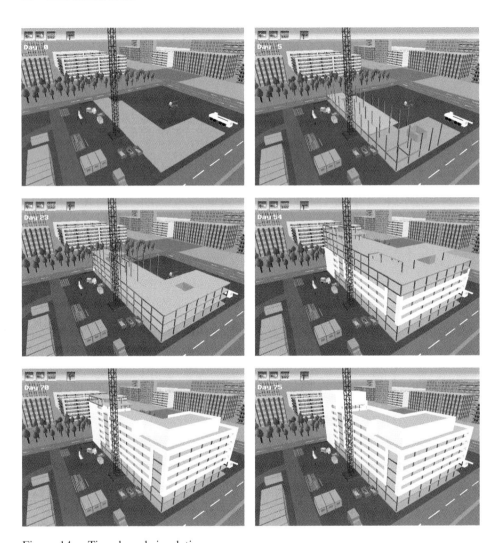

Figure 14.   Time-based simulation.

USING THE MODEL

Once the simulation scenario is built in, the construction process can be simulated day-by-day (see Fig. 14) by pressing the forward or backward arrow icons on the screen (see left corner, first image, of Fig. 14). The user can stop simulation any-where or select to "jump" to any day he/she wants to explore the project at any particular time. The exploration can be done interactively by walking through or by changing views (see Fig. 15).

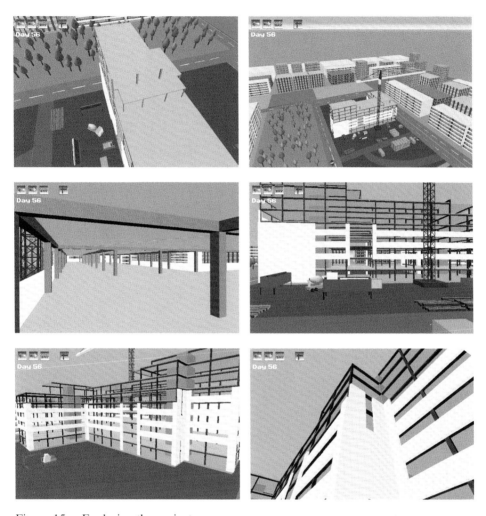

Figure 15.   Exploring the project.

FUTURE DEVELOPMENTS

This application has proved to be an excellent communication means for different purposes, e.g. to run project demonstrations to authorities or other important interest groups, and to run planning meetings participated by different subcontractors. There is still much scope for refining this model. There should be an increase in the amount of activities shown, e.g. internal walls, roofing, windows and the external glass canopies. Future work could focus on making the building site more active and additional plant machinery could work away in the background. This has already been experimented on an another project. On the YIT Project, the

Figure 16.   Monitoring the project (superimposed manually).

plant machinery was made controllable rather than self-automated. Yet, it should be possible to combine the two ideas, so that the machines will be computer controlled, but can be taken over by the user at any time.

The materials placed on the building site can also be simulated accurately, if storage area data was made available as a function of time. Materials could be delivered to a certain location, on a certain day, and then disappear as the building takes shape.

It is also possible to use the model during the construction stage for progress monitoring purposes. If remote access to the site is available (as described and demonstrated in Retik et al., 1997), real as-built pictures from site could be superimposed with VR as-planned pictures as demonstrated in Figure 16, so the project progress could be judged visually.

BENEFIT ANALYSIS

Construction production planning aims to find out the means to achieve and maintain the most efficient site production. The site engineer needs to consider various options and choices, and do trade-off analyses between contradictory targets like keeping minimum buffer between jobs and allocating same time and adequate space for each work. 4D models can help the decision-making by offering support to production planning challenges as explained in Table 1.

The results presented in Table 1 are based on site managers' and engineers' estimates. In this survey the site managers and engineers from the THK case project

Table 1.   Usage and benefits of 4D modeling in building construction.

Challenges/questions	4D model usage	Results
Are the schedules realistic and feasible?	Easy to explain to architect, structural and service engineers when the designs and drawings are needed. Easy to show to decisions makers the dead-lines of decisions concerning design choices.	Efficient procurement with complete designs and drawings. Less waiting in the site. Materials delivered JIT.
How could I more effectively market my project?	Easy to show around the (virtual) real premises to customers earlier. Customers can start designing the interiors sooner.	Premises rented earlier. Less rebuilding due to changes.
How should I allocate resources and how can I avoid production bottlenecks?	Bottlenecks are easier to notice in advance. More efficient resource management.	Less reworking. Less rush at the end of the project. Less unnecessary resources.
Have the safety factors on site considered properly?	Dangerous places easier to find out. Better-organized site.	Safer site. Better working conditions. More satisfied workers.
What are the appropriate tower crane size, location and capacity?	Tower crane size is based on the actual loads and locations of the loads.	Right tower crane size, location and capacity.
How can I brighten up the image of my company?	Taking advantage of novel technology.	Image of the front-line company.
How can I more efficiently analyze the logistics of the site?	Right locations for storage areas easier to detect. Material flows can be analyzed in VR.	Efficient material handling. Field factories located correctly. Storage places located correctly.

were familiar with the 4D model that was developed for their project. Based on that knowledge they estimated the possibilities and potential benefits when a fully operational 4D system would be used in a project comparable to THK building project. Furthermore, this survey covered the magnitudes of the potential benefits. The potential benefits were estimated using ranges (minimum, most likely, maximum) and possible correlation between benefit items identified. After this, the overall benefits were analyzed using @RISK software package. Figure 17 and Table 2 present the summary of the results of the benefit analysis. The results demonstrate that one can expect considerable advantages from the use of operational 4D system in building construction. It seems that the biggest potential for benefits lies in the reliability of the schedule, bottleneck identification and resource management. The results are in line with Akinci et al. (1997).

exists to help subcontractors manage their resources beyond the productivity improvement literature for individual activities. There is a complete lack of tools that relate a subcontractor's resource allocation across projects to cost (O'Brien & Fischer, 2000). Costing methods follow the assumptions of network scheduling methods, focusing on costs of individual activities and of single projects. No allowance is made for interaction between projects. Yet subcontractors emphatically do manage their resources from a multi-project perspective, taking a variety of actions to optimize resource allocation across projects in the face of changing conditions.

Costs to subcontractors in a multi-project context occur because of changes in project schedule and scope. If changes did not occur, subcontractors and suppliers could predictably allocate their capacity (resources) to projects and they would not suffer any capacity related costs. (Further, bidding would ensure an efficient allocation of capacity across firms (see O'Brien et al., 1995 for further discussion)). However, changes in schedule are a common occurrence on construction projects. The causes of such changes are numerous and well cataloged in the construction literature: weather, owner directed changes, problems with permitting, unexpected soil conditions, materials delays, accelerations, rework that affects schedule, coordination difficulties, etc.

In one sense, existing 3D and 4D CAD tools are an attempt to control the incidence of changes on construction sites. 3D/4D tools improve designs, reduce the incidence of time and space conflicts on construction sites, improve the materials flow, and improve schedule reliability. Thus 3D/4D tools may reduce the number of changes on project sites and therefore improve subcontractors' ability to plan their resource allocation across projects. But it is the nature of projects to be dynamic; while generating one set of improved capabilities, 3D/4D tools do not remove the changes in project schedule and scope from influences such as weather. Moreover, the capability of 3D/4D tools in design may make owners more rather then less likely to initiate changes, particularly in a business environment that demands speed and flexibility.

Changes to projects are given, and thus subcontractors will continue to face multi-project resource management challenges. Unfortunately, existing 3D/4D tools lack a multi-project perspective, having been developed in a single project context. These tools do not aid a subcontractor in its resource management and may in some cases hinder it. For example, many 4D models make assumptions that resources are fixed or that there is costless flexibility in assigning resources. Other 4D models that automate schedules assume that methods direct resources, and assign subcontractor resources based on methods and schedule needs. Yet for many subcontractors, it is the availability of resources that influences the choice of methods and staffing needs on a job site. Thus the real-world considerations of subcontractors may be in direct conflict with the virtual dictates of a 4D model.

This paper discusses extensions of 3D/4D modeling methods to subcontractors' multi-project resource management needs. A case study of the progressive subcontractor Pacific Contracting provides details of the constraints and consider-ations a subcontractor makes when making resource allocation decisions in a multi-project environment. From the case findings, a conceptual basis for multi-project resource management and costing is presented, allowing definition of several questions pertinent to subcontractor resource management. These provide a foun-dation for examination of current capabilities in 3D/4D CAD and discussion of extensions. In particular, there needs to be development of 4D tools that automat-ically assess the affects of resource re-allocations on project performance and ripple effects across projects.

PACIFIC CONTRACTING PART ONE: RESOURCE MANAGEMENT PROBLEMS

The Pacific Contracting case study presented in this section is an intimate descrip-tion of how a subcontractor manages its business, particularly with regard to man-agement policies in relation to site conditions and resource allocation. Such a view of subcontractors is largely absent from the construction literature. One explana-tion for this lack of literature about subcontractors is the arm's length relationships that exist between subcontractors and general contractors. Subcontractors are there to provide a contractual service. Hence, the extant literature on procurement and contracting serves as a substitute for a more specific literature about subcon-tractors. Discussion of the relative merits of contracts and incentives can be found in Abu-Hijleh & Ibbs (1989), Ashley & Workman (1986), Griffis & Butler (1988), Stukhart (1984) and Uher (1991). This literature does provide a useful way to structure and understand relationships and, of course, contracts are a necessary aspect of business practice. However, none of this literature explicitly considers the production choices facing subcontractors or their internal cost drivers.

Some literature provides a macro-level view of the conditions facing sub-contractors (Gray & Flanagan, 1989; Bennett & Ferry, 1990) or focuses on the contractor–subcontractor relationship (Hinze & Tracey, 1994) rather than on internal subcontractor policies. While none of this literature details resource management policies at the operations level, there are several points of agreement between the literature and the Pacific Contracting case study that suggests Pacific Contracting is a common example of its genre. One point common among all the authors is that subcontractors are commonly subject to changing conditions with poor man-agement by general contractors. In a review of subcontracting in the United Kingdom, Bennett & Ferry (1990: 271) note that:

> … under construction management contracts, and to a greater extent under management contracting, the specialists are just thrown together and told to sort things out by themselves.

significant advance in the development of construction modeling technology. The ability to construct a virtual prototype before field construction began dramatically decreased interference among major systems (e.g. piping, mechanical, and structural systems) and increased the speed and quality of design review. Other early uses of 3D CAD models included limited studies in constructability. On one project, 3D CAD models and renderings were used in coordination meetings to discuss trade sequencing (Griffis et al., 1990). Research activities associated with that project explored the link between traditional simulation of construction activities and simulation of construction in the context of the 3D model (Griffis et al., 1991). This work quickly led to the development of 4D CAD approaches to modeling both the facility and the construction process. Early 4D CAD efforts manually integrated schedule information with 3D models to represent the planned state of facility construction at fixed points in time. Later 4D CAD research efforts automated the link between scheduling software and 3D models to create more flexible and dynamic 4D representations of construction progress. Research has also focused on generation of automated (4D) schedules by reasoning about the 3D design and construction process information (e.g. Darwiche et al., 1988; Thabet & Beliveau, 1994, 1997). Current application and research frontiers in 4D CAD include detailed work planning (Riley, 2000) and coordination of multiple trades in a dynamic and uncertain project environment (Akinci & Fischer, 1998; Tommelein, 2000).

Research has also begun to incorporate construction costs in 4D CAD models. Staub-French & Fischer (1999) review the practical needs of cost, schedule and scope integration and outline an approach to cost planning at the activity and object level in a 4D environment. Staub-French & Fischer's (1999) research perspective can be shown in Figure 6, where cost is a third link in a triangle integrating cost,

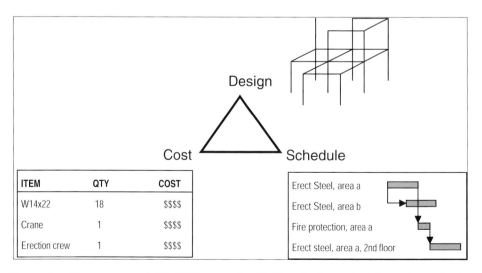

Figure 6.   Current state of 4D CAD research—linking design, schedule, and cost in the context of a single project.

scope (design), and schedule in the context of a single project. They identify several impediments to effective integration of project information, in particular different levels of aggregation in the use and generation of design, schedule, and cost data. Through case examples, Staub-French and Fischer also show how the standard construction accounting and control methods and existing software are often too rigid to accommodate dynamic reasoning about and representation of construction methods. Building on the work of Fischer & Aalami (1996), they propose a conceptual schema that will accommodate both different uses and different levels of aggregation of construction information in a unified format that will allow integration of cost, schedule, and scope on a project.

While creating increasingly more powerful tools, research in 3D/4D CAD has stemmed from a single project perspective. Existing 3D/4D tools do not directly support a multi-project viewpoint (a perspective this author has called "5D CAD" in to distinguish it from single project approaches (O'Brien, 2000b)). Based on the discussion above, a multi-project or 5D CAD tool should support decisions about cost, time, and resources at the firm level. While more research needs to be performed developing integrated costing models to provide a decision support framework for multi-project resource management, several extensions to 4D techniques are possible. Let us consider them in the context of the resource management questions presented above. Consider the first three questions:

- Which project should we borrow resources from?
- Which project should we deploy idle resources to?
- What are the ripple effects of a resource re-allocation and what do they cost?

These are intimately related and concern the daily operational decisions a subcontractor must make about where and how to deploy resources. Should there be a problem or acceleration on one project, the first question asks which of the subcontractor's other projects should resources be borrowed from. This is not a simple question as moving resources from one project to another may simply solve one crisis by creating another. Many subcontractors do shift resources, and many subcontractors do seem to always be running late. Similarly, deploying idle resources to a project may not help that project. If the subcontractor completes work early on one project, which project should it allocate those now idle resources to? Ideally, the subcontractor will maintain a level use of resources (perhaps with some spare capacity from a queuing perspective (Hopp & Spearman, 2000)), and should there be changes, the subcontractor will seek to minimize the cost or consequences of any ripple effects.

Thus, for the first three questions, what is needed is an extension of 4D CAD that allows the ripple effects of resource re-allocations to be modeled. This can be accomplished manually today insofar as we can assess productivity on a project for a given resource level. With some knowledge of the overall schedule for each project (especially with regard to float and space) and any penalties for delays (and incentives for early completion), it is possible to model the impact of a given resource re-allocation on each project. The overall impact of any proposed change can thus be assessed by individually evaluating each affected project. Of course, this is extremely cumbersome to do manually, and should a resource re-allocation generate significant further re-allocations (i.e. large ripples) it may be impossible to manually enumerate all the possibilities. What is needed is a 4D tool that allows *automatic* exploration of the impact of resource re-allocations on a project. With this, we also need a tool that tracks the impact across projects (including ripple effects), allowing exploration and evaluation of changes. This appears to be a significant research challenge, although some groundwork has been put in place by the automatic space and conflict evaluation work of Akinci & Fischer (1998) and Thabet & Beliveau (1997), and the resource tracking work of Choo et al. (1999).

The second set of questions concern themselves more with planning than with operations:

- How do we value the flexibility of resource deployment on a project when bidding?
- What is the most profitable mix of projects?
- How does production technology affect our ability to accommodate changes?

These questions are seemingly unrelated, but consider that a subcontractor's management knows there is a high probability of resource re-allocation. In this case, the management would like to have a set of projects where re-allocation of resources is not costly. Conceptually, one set of projects will be more or less flexible in accommodating changes than another set of projects. This flexibility directly relates to profitability, and thus subcontractor management would like to assess the value resource flexibility on projects both individually and as a set. To a certain extent, some subcontractors already practice choice about project mix; e.g. one steel fabrication and erection subcontractor studied likes to work on one large project and several small ones at any given time (O'Brien, 1998). Such a mix provides the firm with flexibility in meeting changes while keeping near full use of its productive resources.

Valuing flexibility is not simple. Influences on the cost of shifting resources include not just schedules and any associated penalties, but also the technologies involved. In a simple sense, a subcontractor cannot shift from one technology to another if there are different equipment and skills involved. In a broader sense, construction technologies can accommodate changes in resource level with different levels of impact on productivity (see O'Brien, 1998, 2000a for more discussion and examples). With knowledge of the impact of a technology on productivity, conceptually it is possible to extend the 4D/ripple-effect technologies envisioned above to provide detailed knowledge about the value of flexibility for a given set of projects. Essentially, the question is the same: what is the impact of a change? Of course, in the early planning stages less detail will be known about the project(s) than in the operational stages. It is unclear if the same set of technologies that will allow detailed assessment and enumeration of changes for improvement in operations will work in a less information rich environment. It is likely that the data from the 4D technologies envisioned would have to be put into some sort of options framework (Trigeorgis, 1996; Brennan & Trigeorgis, 2000) that can accommodate uncertainty.

The final questions directly concern short-term operational planning rather than accommodation of changes after they occur (questions 1–3) or with pre-bid or investment planning (questions 4–6):

- How should we negotiate a schedule with a contractor?
- How much reserve capacity should we have to accommodate changes?

If the first three questions (above) concern themselves of what to do after a change occurs, these final questions concern themselves with planning for those changes. With some knowledge of the value of a mix of projects and the influence of schedules and incentives on that value, subcontractors can better address the detailed negotiations with contractors about cost/price, schedule, materials, etc. Similarly, with a solid plan of projected resource assignment to projects over time, subcontractors can better negotiated a detailed schedule of work area hand over dates on individual projects. With regard to extensions to 4D tools, it is likely that a combination of the tools described above will provide subcontractors with the requisite knowledge about what to do. New development that is needed is a better visualization tool to support negotiations between the contractor and the subcontractor(s).

As for schedule changes, improved knowledge about the impact of possible changes on cost and schedule can guide the subcontractor in making choices about reserve capacity. Subcontractors can choose to keep some resources in reserve for a given mix of projects (type and amount of resources determined by the extended 4D tools), or they can choose to make strategic investments in new resources (either rental or new permanent investment). As with schedule negotiation, some combination of the tools envisioned above will provide the necessary decision support.

CONCLUSIONS

Methods and models in 4D CAD have many benefits to construction projects, not least an ability to resolve design and construction conflicts before they occur in the field. Efforts in 4D CAD can be seen as a way to reduce the incidence of costly changes on construction projects. However, even in a world with widespread use of 4D CAD, changes in schedule and scope will remain a salient feature of construction projects. A core competency of subcontractors is the ability to adjust to these changes in a low cost manner. The principal difficulty that subcontractors have in adjusting to changes is maintaining a level use of their finite resources. Any change

Inputs to planning

Space planning may be considered as a technique to evaluate scheduling or sequencing alternatives to determine if spatial conflicts exist between different trades. Judgment must then be made as to the severity of potential spatial conflicts and potential course of action. For this reason, it is assumed that a 4D model of the facility to be analyzed is a required input to the type of detailed planning described in this paper. This 4D model consists of a 3D product model of the facility in which all objects to be planned are related to activities in a CPM construction schedule.

Automated input: elements of 4D space planning that may be automated

- *A 3D model of each work area to be considered in the planning process.* Ideally, the development of a 3D product model should include the modeling of work spaces, in appropriate size, for each object in the model. An object oriented 3D modeling environment would permit these work spaces to be predefined for various components of a building and generated concurrently with the 3D model. Figure 7 illustrates various work areas that could be associated with product model objects for *unit* work spaces in isolated locations (A), *overhead* sections of ductwork, pipe, or cable tray (B), *linear* wall assemblies, in-wall plumbing, and electrical or wall finishes (C), or *vertical* sections of pipe risers, elevator shafts, and ductwork, etc. (D).
- *A property database for work spaces and associated* spaces. The inherited properties of construction work spaces described earlier must be included as attributes of work space objects in the model.

User input: elements of 4D space planning required by user

- A sequence in which model objects that are associated with unique construction activities become active would need to be determined by the planner. For example, a single schedule activity: "Install 4th floor light fixtures" requires further detail to define the order in which fixtures are installed. Subsequent duration adjustments to schedule activities would then adjust the rate at which fixtures would be installed and the rate at which a crew would move through the work spaces for that activity. It should also be noted that this sequence would most likely be changed to adjust work direction and flow rates.
- Assigned positions of material access points and storage areas for discrete or multiple work areas. These positions are project specific, and cannot be inferred from the position of objects in the product model. From these locations the distance and proximity relationships of material paths between access, storage, and work spaces may also be calculated automatically. It would be advantageous to define access points for material loading and waste removal, and allow paths between these points and storage locations to be inferred by a planning tool.

- Lead times or fixed dates for material delivery to storage spaces which permit the timing of material movement and required storage space to be calculated from the CPM schedule information of respective objects in the product model. For example, light fixtures may be loaded onto building floors two weeks prior to installation. Using lead times permits material loading dates to shift along with schedule changes, while fixed dates might be determined by manufacturing or logistical constraints on specialty materials.

Planning outputs

The ultimate product of 4D modeling and space planning should be a construction plan that is free from disruptive spatial conflicts. The automated detection of potential conflicts between work space, storage areas, and paths of different crews represents the primary goal of the 4D modeling process because it permits complex and long duration work sequences to be evaluated and re-evaluated after adjustments are made.

Additional benefit would also be observed if automated reasoning about the severity of potential conflicts were possible. For example, a planner may choose to evaluate only specific types of conflicts or only those that occur for a planning interval of one week or more. As discussed earlier when the concept of density was introduced as an attribute of 4D objects, it might be possible for particular activity spaces to occupy the same location concurrently with little or no negative impact to production. Some materials may be stacked and some paths of low density may be shared by more than one activity. Rules for evaluating the severity of detected spatial interferences based on their durations and density would also be beneficial, and is the subject of current research.

Table 2 identifies six types of spatial conflicts that would be beneficial to detect, and a range of suggested densities that may be used to assess the severity of the conflict. A "full" density range indicates that the type of conflict should be identified and resolved at all density levels of the spaces involved. A "variable" density range indicates that a conflict will only be identified between those types of spaces if the sum of the densities of those spaces exceeds an unacceptable level.

Table 2.   Potential spatial conflicts between $Activity_a$ and $Activity_b$.

Conflict type	Represents conflict between	Density range
(1)	(2)	(3)
$Work_a$–$Work_b$	$Activity_a$ and $Activity_b$ work spaces	Full
$Storage_a$–$Work_b$	$Activity_a$ storage and $Activity_b$ work space	Full
$Path_a$–$Work_b$	$Activity_a$ path and $Activity_b$ work space	Variable
$Storage_a$–$Storage_b$	$Activity_a$ and $Activity_b$ storage areas	Variable
$Path_a$–$Path_b$	$Activity_a$ path and $Activity_b$ path	Variable
$Path_a$–$Storage_b$	$Activity_a$ path and $Activity_b$ storage area	Variable

Tommelein, I.D. & Zouein, P.P. 1993. Interactive dynamic layout planning. *Journal of Construction Engineering and Management* 119(2): 266–287. New York: ASCE.

Thabet, W.Y. & Beliveau, Y.J. 1993. A model to quantify work space availability for space constrained scheduling within a CAD environment. *Proceedings of the 5th international conference computer civil and building engineering: 110–116.* New York: ASCE.

Zouein, P.P. 1995. *MoveSchedule: a planning tool for scheduling space use on construction sites.* Ph.D. Thesis, Civil and Environmental Engineering Department, University of Michigan.

THE LINK BETWEEN DESIGN AND PROCESS: DYNAMIC PROCESS SIMULATION MODELS OF CONSTRUCTION ACTIVITIES

E. Sarah Slaughter

MOCA Systems, Newton, MA, USA

Abstract

Recent research at MIT has developed a theoretical framework and specific methodologies, resulting in computer-based process simulation models for 12 selected construction processes, to systematically assess the construction process impacts of design and technology alternatives. Specifically, the research takes two distinct approaches. First, the research allows the explicit linking of the design process to the construction processes through the focus on the specific components and subsystems in the built facility. The second approach considers the system and inter-system relationships, both spatially and operationally, throughout the construction phases. The combination of these two approaches provides a means through which the impacts of particular designs and processes can be analyzed with respect to a specific system and for the project as whole, considering the primary, secondary and tertiary impacts. The research can have significant implications for improving the efficiency of the construction of facilities and the performance of these completed assets. In particular, the application of the methodologies developed in this research can improve the robustness of facility design and technology selections through the explicit evaluation of multiple alternatives for a specific project and its objectives. It can also provide a common basis of analysis for design and construction organizations, to collaborate and reconcile design and construction objectives.

Keywords: dynamic process simulation, design/technology innovation, assessment of design and process

INTRODUCTION

While many people refer nostalgically to the days of the masterbuilder, when a single person directed the design and construction of great human works, others dedicated to a specific set of activities. In addition, unlike the continuous process-ing that is required in chemical plants, construction activities are performed within a project, with a defined beginning and end, to accomplish specific objectives. Each construction project is by its nature unique, consisting of new combinations of components, systems, and resources to create the facility. The industries also differ in their performance measures of the processes. In general, process and manufac-turing industries (and their simulation tools) focus on the process time per unit and overall process throughput (i.e. the rate at which the units are processed through the complete cycle). In contrast, the performance of a construction project is judged by the cost and duration needed to realize the facility design.

This research developed a construction process simulation system to address the shortcomings of existing estimating techniques, and to provide a tool to directly calculate the time, cost, and worker safety impacts of design and technology alter-natives. Taking a different approach than the current research in 4D CAD and queuing-based simulation, this research is strongly complementary to the ongoing efforts in those areas, providing a different view on the link between design and construction, focusing particularly on the assessment of specific alternatives within a particular project. The research built upon the process modeling approach used in the chemical industry, and extended the theoretical framework to accommodate the specific requirements of the construction industry.

RESEARCH APPROACH

Recent research at MIT has developed the theoretical basis and specific methodolo-gies to systematically assess the construction process impacts of design and technol-ogy alternatives. The objectives of the research were to characterize construction processes by system and material, and to assess design and technology alternatives within systems and across systems in the project as a whole for their impacts on duration, cost, and safety (Slaughter & Eraso, 1997; Slaughter, 1999). The specific research methodologies culminated in the creation of specific computer-based dynamic process simulation models that can be used to evaluate specific designs and technologies for particular projects.

This research resulted in the characterization and model development for 12 system and material-specific construction processes. The models are stored in a "library" of system and material-specific processes, and can be accessed and used for any size or type of construction project that employs the process (Table 1). The four general systems are the structure, exterior enclosure, services, and interior finish. The inter-system links are incorporated in a meta-model for the whole proj-ect, with status and information tracking across each of the systems to represent overall progress.

Table 1.   Construction process models completed or in process.

System	Material-specific model
Structure	Steel
	Cast-in-place concrete
	Light wood framing
Exterior enclosure	Precast concrete panels
	Glass/metal curtainwall
Services	HVAC (heating/ventilation/air conditioning)
	Hot water heating
	Plumbing
	Fire protection
	Electrical
Interior finish	Interior walls
	Suspended ceiling

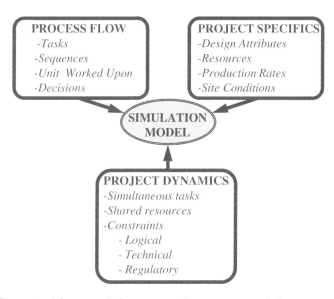

Figure 1.   Conceptual framework for construction process modeling system.

The modeling system captures the common aspects of each construction process (i.e. tasks and associated resources and the related production rates by specific entity) while remaining completely adaptable to the elements that differ between projects (i.e. design attributes, number and character of resources), and the dynamic aspects of each project (e.g. site management strategies) (Fig. 1). Each process model is applicable across all types and designs of projects and can be immediately applicable to a new project or design without recreation or respecification of the tasks and their resources, and the sequence of tasks. It also captures the aspects of the design configuration of the components, subsystems, and systems that affect the process.

Structure of the models

The models are configured to use the detailed design attributes for a specific project as input. The modeling system uses this data to chart the path of each component through each relevant task in the process according to its characteristics, and to map the progress achieved in each subprocess by logical sequence and location to ensure progress for the process as a whole. Status information by subprocess, process and location is also transferred across the system-specific processes to capture the dynamics of the whole project. The output for the model is the duration of each subprocess and process by location, the duration-based costs of the on-site resources, and an index measuring the exposure of workers to dangerous conditions.

The resources are pulled from the pool of resources as each unit progresses through the tasks and the resources become available. It is also possible to set the priority level of the task or set of tasks to establish the distribution order of the resources. Establishing the resources with each task or small set of tasks ensures that relative progress is maintained across all tasks and subprocesses.

The structure of the models is scalable, being able to accurately model small and large construction projects. The meta-model of the whole project is also scalable, and can accommodate either a few or many specific processes. The modeling system is also extendable, since additional processes can be added to the meta-model.

The simulation results are the time taken to perform each task and each sequence of tasks based upon the availability of resources and the specific design. These results can be aggregated by system-specific process and across the project as a whole to directly estimate the project duration. The on-site resource costs, which include the labor and equipment, are calculated directly from these time estimates. The duration-based cost estimates calculate the cost of having those resources on site to perform the tasks. The index of the exposure of workers to dangerous conditions is also directly calculated from the task duration. The OSHA-identified causes of worker injury are matched to the exposure of the workers to each injury cause for each task. The index scales the exposure of the workers to each injury cause by the amount of time workers are performing the tasks (Slaughter & Eraso, 1997).

The modeling system clearly links design and technology alternatives to the details of the processes, to most effectively represent all of the project performance impacts. The models can also incorporate organization-specific knowledge and expertise, and provide a realistic basis in which to evaluate alternatives for new processes.

Methodology

The research methodology consists of two parts. The first portion is the development of a systematic methodology to characterize complex construction processes. The second portion of the methodology is the development of a set of compatible, consistent dynamic process simulation models.

Construction process knowledge is usually not documented in written or pictorial form, but is rather obtained through experience, including hands-on training and participation in numerous projects. Therefore, creating a detailed characterization of a construction process, including identification of each task in its proper sequence with required resources and related production rates could not rely upon existing documentation. The methodology to characterize the construction processes was developed to ensure the accurate representation of the on-site activities. The characterization methodology relied upon actual field data and in-depth interviews with personnel at design and construction companies, similar to the methodology employed in other research (Thomas et al., 1990). The process characterization was then translated into the dynamic process simulation model.

In-depth interviews with personnel in general and specialty contractor companies, as well as designers, owners, and other knowledgeable parties, provided data for the specification of relevant design attributes, common types and quantities of resources throughout the US, and related production rates. They also provided critical expertise in assessing the validity of the completed models. Over 75 companies were involved in this stage of the research, contributing to the initial characterization of the process, and verifying that the process characterization is complete and accurate. In addition, over 100 construction sites were visited to conduct direct field observations (Table 2).

The process characterization was translated into a computer-based process simulation model. The objectives of the model development were to explore the representation of the tasks and activities, including the ease by which the models can be modified to represent a specific project, and to accurately model the design and technology alternatives. During the research at MIT, the simulations were run using commercially available simulation software, SimProcess™, which was developed to model business processes. SimProcess is a hierarchical, discrete event simulation package, which provides basic simulation functions (e.g. gate, split, or join) for easy assembly into subprocesses. However, because this software was primarily developed to represent cyclic processes on standard items, the development of the construction process modeling system required significant modifications of the simulation environment in multiple areas to better represent the distinctive characteristics of construction activities.

Table 2. Construction sites visited (1993–1998).

Type of facility	Number of sites
Institutional	24
Office	29
Residential	21
Retail	10
Industrial	6
Other	9
Total	102

The model results for a specific building design were sent to industry experts to assess the accuracy of the results, and in every case the model results were within 1–5 % of industry duration and cost estimates. The specific system design for each construction process was also defined to reflect a representative building type, and was reviewed by relevant designers. The development of each process simulation model has taken 18–24 months, including the estimation of results for the proto-type building, and calibration to accurately reflect actual project performance.

These results become the baseline against which selected scenarios are compared, to analyze the relative system-level impacts of alternative designs and technologies. The scenarios can also be used to analyze the attributes of specific proposed design and technology innovations, to identify certain characteristics that may increase or decrease the expected benefits and to influence the development of more appro-priate construction innovations. In addition, the scenarios can be used to explore potential complementary aspects of alternatives, when a combination of innova-tions may provide greater benefits than the sum of the individual innovations.

Design and process links
The modeling system provides explicit links between the design and the construction process at many different levels. These levels include the specific components, their subsystem and systems, and the specific configuration of the components and systems spatially and temporally for a particular project. It also provides a common basis in which the design and construction team can plan and coordinate the con-struction activities, and reconcile the design and construction objectives. As clients increase their requirements for high quality facilities obtained at a reasonable price and within shorter durations, design and construction professionals increasingly recognize the incentives to work together.

The design is linked to the process through the type and quantities of the compo-nents, and also through their configuration. For instance, if a pipe run is a straight line between the riser and the fixture, there are fewer connections than a pipe run with multiple bends. As a result, the straight run will take less time to place and connect, which will influence the progress of the process as a whole. In some cases, the progress on that particular location in the building may be critical to the progress for subsequent processes, and the time to place and connect the pipe lengths for the run can directly influence the overall project performance.

The spatial relationships of the design elements also influence the overall time, cost, and worker safety of the project. For example, for a floor of an office building, the restroom facilities could be centralized in one area, or separated into several zones. These design alternatives could be used for one, several, or all floors, and the relative costs and time for each design alternative will differ, depending upon the design alternative selected for each system for each floor and the relationship to the contiguous floors. Since the plumbing must be installed before the interior finish can be placed for those rooms, and the interior finish must be complete before the fixtures are installed, the design layout of the facilities determines the rate at which the restroom facilities are completed for each floor, which in turn determines the completion time for the project as a whole.

LINKED DESIGN AND PROCESS CHANGES: SIMULATION RESULTS

The duration, cost, and safety results are calculated for a specific project, for each design alternative or set of alternatives. Examples of the representation of the links between the design and process are discussed here for several different systems (i.e. structural system, exterior enclosure system, and services system) with the simulation results to demonstrate the applicability of the research approach to actual construction projects.

   The baseline for the analysis is a five-story office building, with a footprint of 30.5 by 38 m (100 by 125 feet), with the floor-to-floor height of 3 m (10 feet) and a bay size of 7.6 by 7.6 m (25 by 25 feet) (Fig. 2). The structural system described here is cast-in-place reinforced concrete columns and beams with a two-way slab of 200 mm (8 inches) (Carr, 1998; Slaughter & Carr, 1999). The exterior enclosure system analyzed here is glass curtainwall system (Attai, 1997; Slaughter, 1997). The service system described here is a domestic plumbing system, including hot and cold potable water with a drain, waste, and vent outflow (Murray, 1999).

Cast-in-place reinforced concrete structure results
Significant on-site resources are required to construct a cast-in-place reinforced concrete structure. While this structural type performs very well under extreme loads (such as seismic conditions), it does take many direct labor hours to fabricate

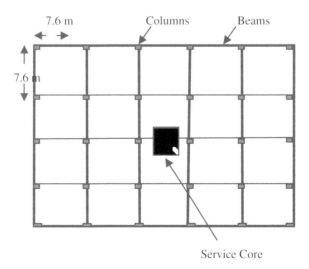

Figure 2.   Floor plan for prototype building.

to a range of different conceptualizations and associated models. This will already be clear to the reader who has reviewed the various papers of this book.

HISTORIC BACKGROUND

4D models are the product of a long evolution in computer modeling. While early computer input and output was type-based, the need for and benefits of providing graphical output soon became apparent. Three-dimensional (3D) viewing was made possible by "Evans and Sutherland (who) demonstrated a head mounted stereo display as early as in 1965" (mentioned in Issa et al., 1999) in order to create an environment of virtual reality. Vector-based "graphical" monitors were created to avoid the jagged lines displayed on low-resolution type- and raster-based monitors. Still facing limited computational power in the 1970s, graphics programmers were challenged by the desire to provide hidden-line elimination, which was necessary for realistic image rendering.

As early as the 1950s when computers became available, algorithms were programmed to solve numeric problems. Pioneers in civil engineering soon performed structural analysis and other decision support as well as automation tasks using computers (Levitt, 1995; iii–v). Likewise in the 1950s, neural nets were conceived of (though not much researched due to lack of funding) and list-processing languages based on the mathematics of rule-based logic were implemented to support non-numeric processing. The 1960s were a prolific time in terms of development of new computing practices. Programmers in the field of Artificial Intelligence (AI) ambitiously developed a "general problem-solver" called GPS (Simon, 1996), object-oriented programming (OOP) languages came about, and the Internet was established. Soon Xerox PARC played a key role in prototyping new ways for people to interface with their computers (e.g. the SmallTalk OOP language and the mouse were invented there). The notion of blackboard systems developed in the late 1970s (Nii, 1986a, b). Their problem-solving ability relied upon distributed, complementary yet competing knowledge sources, and thereby allowed for problem solving to go on simultaneously at different levels of abstraction. Alternative blackboard architectures emerged in the 1980s (Hayes-Roth, 1985; Durfee, 1988). They are the predecessors to agent-based and later web-based systems.

The early 1980s saw the birth of personal computers, which not only made computing more accessible to the masses but also stimulated the development of local area networks to ease communication and data transfer. This started an era for integration of stand-alone programs (e.g. Tommelein, 1995a) so characteristic of our fragmented industry (Howard et al., 1989).

As for computer modeling in architecture/engineering/construction (AEC) practice, large engineering and construction firms such as Bechtel purchased software and hardware capabilities in the mid-1980s in order to develop their proprietary

3D capabilities and then extend it into a WalkThru™ environment, a direct predecessor of today's 4D models. Other companies such as Stone&Webster tailored off-the-shelf computing environments used in the automobile industry (i.e. Catia) to their own needs. Today's models allow for parametric design (e.g. Revit, 2001).

Processors with ever-increasing speeds and better display technologies at increasingly competitive prices have since flooded the market with computing capabilities. Further miniaturization, from desktops to laptops, then on to palmtops and cellular telephone devices, has made computing today truly ubiquitous. The explosion in popularity of the world wide web and OOP languages such as Java, allowing for distributed, collaborative problem solving using applets, has spurred the development of applications that could only be dreamt of a few years ago.

AEC researchers, teachers, and practitioners have barely begun to scratch the surface of what is computationally possible with today's hardware and software. As an industry, we are all too often conservative and focused on individual projects. Project economics using traditional yardsticks have only on occasion been favorable with respect to promoting the use of cutting-edge practices, including innovative management practices or adoption of the latest computer technologies. Today's computer capabilities are rarely—if at all—holding us back from improving, or—better even—radically reinventing our work processes. The benefit/cost ratios for adopting new practices are shifting, thanks to not only current market pricing and availability of hardware and software, but more importantly due to increasing demand by project owners for added complexity and reliable performance of the facilities that are designed and built. Accordingly, spending more time on prototyping parts of an AEC facility (or the entire facility), prior to construction yields advantage internally to the involved organization and may provide the organization with external competitive advantage. It is within this context that 4D modeling plays an important role.

The true power of 4D modeling as compared to 3D CAD is that by extending geometry with time, the opportunity is created for planning. Planning means giving consideration to alternative means and methods, alternative durations of activities, and alternative times and sequences in which to perform them. The planning task that involves space is not an easy one, however. Most resources considered in traditional planning are scalar (e.g. labor and staffing, equipment, materials, money, and time). Depending on the purpose for which the model will be used, space may or may not be represented adequately as a scalar. Space may be abstracted crudely to be linear (e.g. as in line-of-balance methods, see for instance Harris & Ioannou, 1998), two-dimensional (2D) (e.g. as in layout planning, see for instance Tommelein, 1989; Thabet, 1992; Tommelein & Zouein, 1993; and others), and at best to be 3D. Thirteen topological relationships exist between two one-dimensional (1D) intervals, as needed to model time and linear space (Allen, 1984). One can restrict the meaning of existing English words such as "before", "during", "starts", etc. to express those relations. Similarly, 13^2 or 169 topological relations exist between two rectangles (not to mention arbitrary shapes) in 2D space, yet the English

Table 1.   Space requirements for appliances in functional hierarchy.

	w [m]	d [m]	h [m]	U [m²]	L [m²]	C [m²]
Microwave	0.46	0.30	0.40	0.32	0.32	
Conventional oven	0.54	0.6	0.6	0	0.742	0
Wide conventional oven	0.76	0.6	0.6	0	1.024	0
Range & oven	0.54	0.6	0.9	0	0.742	0.371
Wide-range & oven	0.76	0.6	0.9	0	1.024	0.512
Wide-range & grill & oven	1	0.6	0.9	0	1.331	0.666
Range	0.54	0.6	0.1	0	0	0.371
Wide-range	0.76	0.6	0.1	0	0	0.512
Range & grill	0.76	0.6	0.1	0	0	0.512
Dishwasher	0.6	0.6	0.6	0	0.819	0
Sink	0.5	0.6	0.3	0	0.346	0.346
Wide sink	0.6	0.6	0.3	0	0.41	0.41
Double sink A	0.9	0.6	0.3	0	0.602	0.602
Double sink B	0.9	0.6	0.3	0	0.602	0.602
Refrigerator	0.72	0.76	1.64	0.2432	0.973	0.486
Wide refrigerator	0.82	0.76	1.64	0.2752	1.101	0.55
Two-door refrigerator	0.92	0.76	1.64	0.3072	1.229	0.614

Table 2 illustrate such calculations for the first heating set from {(range & oven)(wide-range & oven)(wide-range & grill & oven)}. This calculation reveals that the lower-end choices leave sufficient space, whereas the higher-end choices do not. Further investigation will reveal what set combinations yield acceptable space use and example calculations are shown in the right-most nine columns of Table 2. Only the last three of these columns present sets of appliances that in *any combination* yield a solution.

Admittedly, calculations like these are tedious to perform but this is exactly where computers are most useful. They can perform calculations and keep track of sets, and thereby enable people to focus on other decision-making steps. Note that in this example, the alternative possible configurations can be enumerated by hand should the designer wish to do so. In more complex design cases this will certainly not be the case.

DISCUSSION

The advantage of the postponed-commitment method is that it keeps track of all possible configurations that meet the requirements. No solutions are prematurely rejected. While applying postponed commitment to define the product, the designer-builder should also consider production process issues, such as procurement (e.g. availability and lead times), fabrication, supply, and construction and final

Table 2. Calculation of set properties.

| | Alternative Set 1 | | | | | | | | | | | | | | |
| | Lower bound | | | Upper bound | | | Alternative heating | | | Alternative refrigeration | | | Alternative heat and refrigeration | | |
	U	L	C	U	L	C	U	L	C	U	L	C	U	L	C
Conventional oven															
Wide conventional oven	0.00	0.74	0.37				0.00	1.02	0.51				0.00	1.02	0.51
Range & oven															
Wide-range & oven															
Wide-range & grill & oven	0.00	1.33	0.67	0.00	1.33	0.67				0.00	1.33	0.67			
range															
Wide-range															
Range & grill															
Dishwasher	0.00	0.82	0.00	0.00	0.82	0.00	0.00	0.82	0.00	0.00	0.82	0.00	0.00	0.82	
Sink	0.00	0.35	0.35												
Wide sink															
Double sink A				0.00	0.60	0.60	0.00	0.60	0.60	0.00	0.60	0.60	0.00	0.60	0.60
Double sink B															
Refrigerator	0.24	0.97	0.49										0.24	0.97	0.49
Wide refrigerator										0.28	1.10	0.55			
Two-door refrigerator				0.31	1.23	0.61	0.31	1.23	0.61						
Total	0.24	2.88	1.20	0.31	3.98	1.88	0.31	3.67	1.73	0.28	3.85	1.82	0.24	3.41	1.60
Maximum space available	3.84	7.68	3.84	3.84	7.68	3.84	3.84	7.68	3.84	3.84	7.68	3.84	3.84	7.68	3.84
Remaining space	3.60	4.80	2.64	3.53	3.70	1.96	3.53	4.01	2.11	3.56	3.83	2.02	3.60	4.27	2.24

assembly. After considering these, the product solution sets will likely be further narrowed, but one or several solutions will remain if the problem is solvable. By contrast, should such issues be considered after the design had been locked in to a single product solution, little flexibility would have remained and any additional constraint might then lead to failure and force another cycle of design iteration.

The likelihood for additional considerations being revealed after design "completion" is great. During design, owners will likely learn about and become interested in options they did not think of initially, when they learn about new functional capabilities being desirable and/or available on the market. When choices such as "Which of two double sinks is best?" leaves the owner indifferent, issues such as availability may prevail. Owners may also express different product and process preferences regarding life-cycle concerns at the time the system in its entirety has been configured.

A design that is further refined to include component choices and geometry can be represented in 3D CAD with timing information added to reflect manufacturing, supply, and construction assembly sequencing. Questions to be answered using the 4D model may include:

> Once all parts have been selected and dimensioned, where will they fit within the kitchen space provided?
> Upon delivery and installation, will all appliances fit through the door opening to the kitchen?

Questions to be answered using an extended 4D model may pertain to numerous other issues. Questions to be answered during production planning may include:

> How well can the owner articulate the requirements and preferences?
> Are these likely to change during the design-build process?
> What are the lead times for getting the various appliances and materials to site?
> How will deliveries be made?
> What are the logistics of staging and moving materials about the site?
> What is the availability of skilled labor?
> In what order will the trades (cabinet makers, electricians, plumbers) proceed to build this kitchen?
> How long does it take to install cabinets and shelving, rough-in plumbing; rough-in electrical; install and hook-up each appliance; finish plumbing, finish electrical, finish tile counter tops?
> When is a component or a system hooked up so it can be tested for operation?

Questions to be answered throughout the product-development process may include:

> Does each individual selection meet the owner's specifications?
> What configuration best meets the owner's needs?

The set-based representation, like the one presented here to extend the usefulness of 4D models, enables AEC practitioners to consider alternatives in products while taking process issues into account. It not only yields better solutions, but also avoids needless iteration. Other industries have adopted this approach (e.g. Ward et al., 1995; Sobek et al., 1999).

CONCLUSIONS

This study has presented sources of variability and uncertainty in product as well as process definition. A world in which no variation or uncertainty is recognized typically gets modeled by means of single numbers, often representing expected values. These values are mathematically speaking most likely to occur but in reality have an almost zero likelihood of actually occurring. Singe values reflecting early commitment tend to lead to process iteration. Systems that are subject to variability and uncertainty in terms of processing times exhibit phenomena such as "starvation" that lead to detrimental performance. Models based on averages are unrealistically optimistic. Systems that are based on unique selections at each design step, tend to cause substantial iteration and rework when conflicts are detected. This chapter has therefore argued for the creation and study of 4D models that explicitly represent variability and uncertainty as well as sets of alternatives, referred to as 4D+. Examples illustrated how 4D+ representations make it possible to better manage the production system that supports integrated product and process development.

People's ability and ease with which they can solve problems depends on the representation that is being used. Researchers and practitioners need to expand their conceptualizations of AEC systems so as to allow for integrated product and process development, and then develop representations and problem-solving methods to enable us to really tackle the problems face.

ACKNOWLEDGMENTS

Several ideas presented in this chapter were refined during much-valued discussions with Glenn Ballard, Carlos Formoso, Hyun Jeong Choo, Marcelo Sadonio, Nuno Gil, Cynthia Tsao, Jan Elfving, Nadia Akel, Michael Whelton, and Yong-Woo Kim. Research leading to the development of CADSaPPlan was funded by grant CMS-9622308 from the National Science Foundation (NSF). On-going research on integrated product and process development is being funded by grant SBR-9811052 from NSF. All support is gratefully acknowledged. Any opinions, findings, conclusions, or recommendations expressed in this chapter are those of the author and do not necessarily reflect the views of NSF.

In addition, subcontractors will become more active in the early phases of the design as the architect and engineer develop the specifications and schematics that form the basis for subcontractors' design. The subcontractors' detailed designs will still need to be approved by the architect and engineer through the shop drawing process, but subcontractors will need to develop CAD modeling capabilities to benefit from this involved process. Also, as subcontractors become more active in the design, they will become better able to assist the general contractor in coordinating the work of subcontractors throughout project delivery. (Civil Engineering, May 1999)

LARGE COMPANIES' VOLUME VS. TOTAL INDUSTRY VOLUME

In 1997 there were a total of approximately two million contractors in the United States alone. Of these firms, only 667,089 were not self-employed. These companies include "establishments primarily engaged in the construction of buildings and other structures, heavy construction (except buildings), additions, alterations, reconstruction, installation, and maintenance and repairs" (US Census Bureau). Also included are companies involved in land preparation, demolition, and construction management. Of these, companies with 1–9 employees accounted for roughly 82% (543,753 companies) of all construction companies not self-employed. That means that only about 18% of construction companies with at least one employee, or just over 6% of all construction work is done by companies with 10 or more employees. In fact, only 55,645 companies, or 3% of all construction companies, employ 20 or more people.

Construction put in place in September 1999 was estimated at US $700.1 billion. Of this, US $540.3 billion was performed within the private works sector while US $159.8 billion was generated on public works projects. It could be assumed that companies with more employees are doing more work. This would lead to the conclusion that only 6% of all construction companies account for a significant percentage of construction in place.

While the medium and large firms constitute only 6% of all construction firms, these firms are doing most of the work. It is important to understand the differences between these firms. Among the medium and large firms, there are only a few really large companies. One-third of 1% of all construction companies (5,684) employ more than 100 people and only 77 of those companies employ more than, 1,000 people. Engineering News Record (ENR) annually reports the top 400 contractors based on total revenues of the previous year. The top 400 contractors of 1999 range from US $78.7 million to US $9.7 billion annual revenues and most have at least 100 employees. The total volume produced domestically by those companies in 1998 was approximately US $127 billion. The total value of construction in place in 1998 equaled US $665.446 billion. That means that ENR's

Table 1.   Top 15 contractors for 1999.

1	Bechtel Group Inc., San Francisco, Calif.
2	Fluor Daniel Inc., Irvine, Calif.
3	Kellogg Brown and Root, Houston, Texas
4	CENTEX Construction Group, Dallas, Texas
5	The Turner Corp., New York, NY
6	Foster Wheeler Corp., Clinton, NJ
7	Skanska (USA) Inc., Greenwich, Conn.
8	Peter Kiewit Sons Inc., Omaha, Neb.
9	Gilbane Building Co., Providence, RI
10	Bovis Construction Corp., New York, NY
11	McDermott International Inc., New Orleans, La.
12	Raytheon Engineers and Constructors, Cambridge, Mass.
13	J.A. Jones Inc., Charlotte, NC
14	Jacobs Sverdrup, Pasadena, Calif.
15	Morrison Knudsen Corp., Boise, Idaho
16	Black and Veatch, Kansas City, Mo.
17	PCL Enterprises Inc., Denver, Colo.
18	Structure Tone Inc., New York, NY
19	The Clark Construction Group Inc., Bethesda, Md.
20	The Whiting-Turner Contracting Co., Baltimore, Md.

Source: 1999 ENR top 400 contractors.

top 400 contractors of 1999 (one third of 1% of medium and large firms) generated 19% of the total construction in place in 1998.

The statistics in Table 1 contribute towards further understanding of the impact of large companies in the overall construction industry. These companies, constituting only 6% of the total number of construction companies, will be the testing ground for 4D CAD. The relative small number of companies that have to buy into the 4D CAD idea, and start incorporating it in their pre-construction and construction phases, increases the probability of 4D CAD becoming a valuable tool in the construction workplace.

COMPANIES THAT PERFORM DESIGN-BUILD WORK

The in-house capability of design-build companies is the breeding ground for a future full-bloom 4D CAD. Implementation of 4D CAD will further lessen the communication barriers between owners, architects, and construction managers. The structuring of a design-build company accounts for these barriers by trying to eliminate them. Furthermore, the structure of a design-build company is set in such a way that implementation of 4D CAD fits and further enhances the working relationships within the company. This makes possible the review of the project's schedule while being designed, which is attainable since both teams, the architect and the contractor, work together for the same company.

The possibility of the contracting side of the project reviewing the way it is going to be put together while in the design phase, gives design-build companies an enormous edge over the traditional design-bid-build method of building a facility. Both teams (designers and builders) will be working simultaneously on a project throughout the design phase. The contractor side will see how the building will be put together and at the same time will alert the designer of any necessary changes that need to be made. These changes can be related to the projects schedule or constructability. This process will enable the minimization of field changes and will give the contractor a better view of the schedule before the project has started, and as a result, improve the possibility of a timely delivery of the project.

PERCENTAGE OF LARGE COMPANIES THAT PERFORM DESIGN-BUILD WORK

Most of the top design-build companies listed in Table 2 are among the top firms (according to the ENR's "Top 400 Construction Companies") listed based on total revenue.

Table 2.   Top 20 design-build firms for 1999.

1	Kellogg Brown and Root, Houston, Texas
2	The Turner Corp., New York, NY
3	Bovis Construction Corp., New York, NY
4	Structure Tone Inc., New York, NY
5	Skanska (USA) Inc., Greenwich, Conn.
6	DPR Construction Inc., Redwood City, Calif.
7	Foster Wheeler Corp., Clinton, NJ
8	Gilbane Building Co., Providence, RI
9	Fluor Daniel Inc., Irvine, Calif.
10	Morse Diesel International Inc., New York, NY
11	Chicago Bridge and Iron Co., Plainfield, Ill.
12	Stone and Webster, Boston, Mass.
13	Peter Kiewit Sons Inc., Omaha, Neb.
14	The IT Group, Monroeville, Pa.
15	Marnell Corrao Associates Inc., Las Vegas, Nev.
16	Morrison Knudsen Corp., Boise, Idaho
17	Parsons Corp., Pasadena, Calif.
18	The Haskell Co., Jacksonville, Fla.
19	Ryan Cos. US Inc., Minneapolis, Minn.
20	The Austin Co., Cleveland, Ohio

Source: ENR's 1999 top 100 design build firms.

As mentioned earlier, the key in making 4D CAD a viable tool for construction management will be the initial ability of this new technique to infiltrate through the "old ways" of management techniques and become a popular and efficient way of overlooking and managing a construction project.

The best chance 4D CAD has in becoming the new tool in construction management is to start introducing itself to the large companies first. It is these companies that can absorb the initial cost and adjustment that comes as a result of implementing this new technique for the first time. The nature of the projects these companies are involved with provides a good ground for the testing of 4D CAD. These projects are large in size and usually innovative management techniques is what determines the timely and "within budget" delivery of them.

The introduction of 4D CAD in these types of projects performed by the large design-build companies will improve the communications among the different levels of parties involved in the project. The obvious result will be a better understanding by the owner of how the future facility will progress through the building phase. In addition, 4D CAD implementation in project delivery will further improve the flow of communication between the design and construction personnel within the design-build company. The 4D technology will prove more beneficial if it is implemented further on the job site by making the 4D model available to the project manager, field engineer, and foreman.

When starting to analyze the numbers, the percentage of top design-build firms that are part of the overall top construction firms is relatively high. These companies are the ones that cover most of the large project design-build work in the country. For 4D CAD this means that only a small percentage of the design-build companies have to implement it in their system of project design and management. This relative small percentage of companies will be the testing ground for the new tool, but at the same time this small percentage accounts for most of the large design-build projects currently being built. This way the 4D CAD implementation will be realistically an attainable goal, since only a relatively small number of companies that will use it will cover the majority of the testing ground (the large design-build projects).

COMPANIES THAT PERFORM CM-AT-RISK WORK

The sixth largest firm in 1999 was Foster Wheeler Corporation of Clinton, New Jersey, producing US $3.072 billion in revenue. Nearly 72% (US $2.204 billion) of its revenues were generated abroad. Foster Wheeler is another company that specializes in industrial projects. They completed 79 industrial projects in 1998. Of those, 38% were performed CM-at-risk. Industries served by Foster Wheeler include oil and gas field development, chemicals, petrochemicals and polymers, pharmaceuticals and fine chemicals, petroleum processing, power generation, cogeneration, and resource recovery.

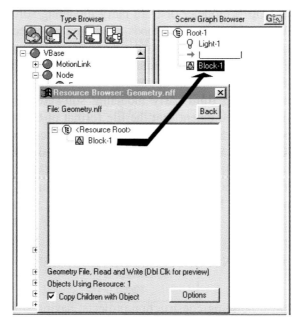

Figure 3.   Object reuse using resource browser of World Up™.

SIMULATING THE CONSTRUCTION PROCESS

Adjei-Kumi & Retik (1997) noted that present planning systems, which use CAD technology to generate and simulate the process of constructing a facility graphically, only go as far as the visualization of project schedules at the component level. They proposed to visualize the construction components and processes in Provysis. That visualization of construction processes, however, does not mimic the real processes of a construction project. Provysis only fades in the construction component assumed to be constructed and fades out the successor components, instead of showing how the object is transported and installed.

Bick et al. (1998) developed a visualization system in the manufacturing domain. In this system, the simulation supports object translation from one location to another location in a specified time. The translation of the object is determined by using origin and destination of object location while time control is achieved by specifying when the process takes place, t_{start} and t_{end}.

In the real world, location and time are the most important variables to represent a dynamic condition. Changing the value of these variables determines where the object will move in terms of location and how long the movement is in terms of time. These two variables are also important for the purpose of simulating the virtually real construction process. The most general construction process is

Distance Phase 1 (D1) = $((X_1-X_0)^2 + (Y_1-Y_0)^2 + (Z_1-Z_0)^2)^{0.5}$

Distance Phase 2 (D2) = $((X_2-X_1)^2 + (Y_2-Y_1)^2 + (Z_2-Z_1)^2)^{0.5}$

Total Distance (DTotal) = D1 + D2

Duration of Transporting = Total = ObjCtr

Phase 1 Duration (TD1) = (D1/DTotal) * ObjCtr

Phase 2 Duration (TD2) = (D2/DTotal) * ObjCtr

Figure 4.　Linear distribution of translation duration.

translation and rotation of an object. Other processes such as installing and removing an object can be derived from the concept of object translation and rotation. For example, installing a pre-fabricated wall can be simulated as translation from the stock location to the destination location.

In order to transport an object from the origin location to the destination location in a specified time, four user variables—origin of location, destination of location, starting time to move, and duration of translating—must be defined. Origin of location and destination of location are user-defined properties that can be represented as x_0, y_0, z_0, xt, yt, and zt. The object will move by changing the value of its translation properties from x_0, y_0, z_0 to xt, yt, and zt. The duration of transporting is used to display a smooth movement of the object instead of leaping from one location to another location.

A more complex translation can be achieved by increasing the number of location user-defined properties, such as origin of location (x_0, y_0, z_0), middle of location (x_1, y_1, z_1), and destination of location (x_2, y_2, z_2). This method creates two phases of movement, from the origin to the middle (phase one) and the middle to the destination (phase two). Since this method creates two phases of movement, the duration of object translation must be distributed linearly in order to make a constant speed of translation (Fig. 4).

COUNTERING SYSTEM AS A SIMULATION ENGINE

In World Up™, the location of an object in the virtual world can be determined by setting the value of the translation property. Adjei-Kumi & Retik (1997) noted that the main purpose of Provysis' process data, such as start times, finish times, duration, and graphical images, is to facilitate the visualization of the simulation of the generated construction schedule. They connected the VR application with project-management tool, Primavera Project Planner. This system supports the usual practice of scheduling in the industry. But it does not explain how to control the time scale. Time scale factor is important to adjust a speed level so that user can analyze the VR simulation at a desired and controllable speed. If the simulation is too fast, it may cause dizziness to the user and it will be difficult to observe the simulation, while a slow simulation may cause impatience.

In order to represent time in the World Up™ universe, a counter system is used. When the simulation is run, World Up™ goes through the entire simulation loop for each frame that the simulation is run (Fig. 5). The counter works every time World Up™ simulates the universe (Table 1). The counter is increased every time that a new cycle of simulation is started. Since the counter will be increased every frame, an object can be designed to translate whenever the object start

Figure 5.   Simulation cycle of World Up™. Source: *World Up™ user's guide* (1997).

Table 1.   Scheduling comparison between real practice of construction and VR simulation.

Activity	Real practice	VR simulation
Time control	Calendar	Counter
Task started	As specified date, i.e. 1 January 2000	As specified counter, i.e. when the universe counter reach 200
Task completion	Number of days, i.e. 5 days	Number of counter, i.e. 100

counter (OSC), a user-defined property, is equal with the counter for a certain duration of ObjCtr, a user-defined property (Fig. 6). For example, an object will be triggered to translate or move, when the counter is the same as OSC, 300, for duration of 100 (ObjCtr). So, the object will stop at its last position when the counter is 400.

The increment of the counter determines how fast the object moves. This increment can be determined by creating an acceleration factor (Acc F) variable, which is a time scale (Fig. 6). The greater the factor, the faster is the translation of the object. This time scale factor influences only the speed of counter calculation, not the number of frames per second at which the universe is rendered, which depends on the computation capability of the computer, number of polygons rendered, number of pixel filled, and the display card. The faster the frame rate is rendered, the better quality of graphic animation is performed.

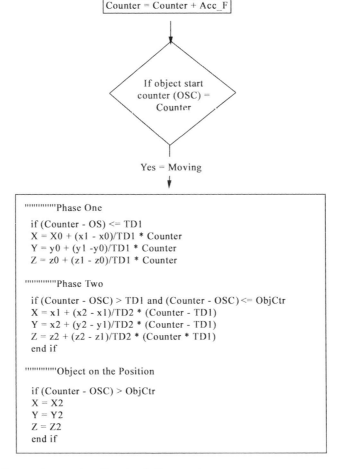

Figure 6.   Countering system for simulation engine.

DESIGN-FOR-SAFETY-PROCESS

General Electric developed guidelines for product design in the 1960s, but a significant benefit was not realized until systematic Design for Assembly (DFA) was introduced in the 1970s. Other design for "somethings"; such as manufacturability, inspectability, and quality; were developed in the 1980s. Then, in the 1990s, Design for X-Ability (DFX) has been used as an umbrella for all of these terms. The potential of DFX is enhanced by the availability of a powerful representation tool such as VR, where design can be represented in 3D graphical data and a walkthrough function enabling the user to discuss any aspect of the object in a virtually real location.

VR of components and processes of a construction project will be used to develop a DFSP methodology, which aims to point out the safety hazards inherited by the construction components and activities. The main idea of DFSP is derived from DFX. The DFX aims to design a product from many viewpoints or characteristics (Gutwald in Prasad, 1996) with the following benefits: achieving a product exhibiting better qualities of X, i.e. Design for Manufacturability (DFM), DFA, Design for Disassembly, Design for Quality and Design for Environment.

EARLY FAILURE DETECTION

Since DFX is an umbrella, the methodology development depends on the specific domain of x-ability developed. For example, DFM is designed by utilizing theories, such as the Taguchi method, and design axioms. However, the most important consideration is that DFX must work as a guideline to evaluate the design. This can be achieved by developing the DFX as an online or offline purpose (Huang, 1996). An online DFX checks its data/knowledge base to ensure that the design decision being considered will not violate the DFX rules. On the other hand, an offline DFX evaluates design decisions after they are made. In this research, an online DFSP is chosen. The safety process will be designed by utilizing a data/knowledge base compiled from theory and axioms. An axiom is a proposition, which is assumed to be true without proof for the sake of studying the consequences that follow from it and the axiom cannot be proven, rather it must be assumed to be true until a violation or counter-example can be found (Sush et al., 1978). In this research, the axiom will be compiled from safety regulations and safety best practice.

Several theories of accident causation exist such as accident proneness, goals-freedom alertness, adjustment-stress, unconscious motivation, situational theories, domino theory, and epidemiological theories (Hale & Hale, 1972; Petersen, 1984). Adams' domino theory of accident causation developed from Heinrich is considered the most relevant accident theory to enable safety hazard recognition. The main reason is that Adams' theory of accident causation discusses the tactical error, a tangible approach, which contains the unsafe conditions as follows (Adams, 1976):

- improper design, construction or layout;
- decayed, aged, worn, frayed or cracked;
- unnecessarily slippery, rough, sharp-edged or sharp-cornered;
- unsafely stored or piled tools or materials;
- poor housekeeping or congestion;
- unsafe established procedure, inadequate job planning, improper equipment provided;
- inadequate aisle space;
- improperly guarded;
- improper illumination;
- improper ventilation;
- personal protective equipment not adequate or not available.

CONCLUSION

The traditional approach using 2D drawings results in bulky and separated design drawings and philosophies. This problem leads to difficulties for the end-user in order to create a 3D mental picture for construction purposes. VR can integrate this design information into one universe and also add in design content such as the construction process, method, and safety planning.

The reusability concept is important in developing a VR universe since it can minimize development time. Reusability can be achieved by using the concept of object-orientation; class & object, and inheritance. Inheritance can be created by applying the concept of generalization–specialization which is also supported by World Up™.

The two important variables, location and time, are identified for the purpose of construction process simulation. In order to translate the object, the location variable must be defined as the origin of location and the destination of location. The time for object translation can be distributed linearly if the location consists of three points or more, such as (x_0, y_0, z_0), (x_1, y_1, z_1), and (x_2, y_2, z_2) in order to obtain a constant object translation.

Since time must also be represented for the simulation purpose, a counter system, counter = counter + Acc_F, is proposed which can simulate a time standard, such as a calendar, in real life. The universe counter will be increased by the factor of "Acc_F" (acceleration factor), which also functions as a time scale to adjust the speed of the construction process as desired by the user every time a frame is calculated by World Up™. Adjustable speed of simulation is important for user friendliness. If the simulation is too fast, the user might get dizzy and might not be able to analyze the simulation properly, but a slow simulation might cause user's impatience.

An example of the use of the model is shown in Appendix A. The model shows a typical 40 story Hong Kong Housing Authority residential block and the model has been used to conduct a safety risk assessment for such units. The walkthrough was found to be realistic and the hazards identified were automatically logged in the risk assessment.

REFERENCES

Adams, E.E. 1976. *Accident causation within the management system.* Professional Safety, October.

Adjei-Kumi, T. & Retik, A. 1997. A library-based 4D visualisation of construction processes. *IEEE conference on information visualization, 27–29 August: 315–321.* ISBN: 0-8186-8076-8.

Bick, B., Kampker, M., Starke, G. & Weyrich, M. 1998. Realistic 3D-visualisation of manufacturing systems based on data of a discrete event simulation. *Proceedings of the 24th annual conference of the IEEE, Industrial Electronics Society: 2543–2548.*

Coad, P. & Yourdon, E. 1991. *Object-oriented analysis,* 2nd edition. New Jersey: Yourdon Press. ISBN 0-13-629981-4.

Faraj, I. & Alshawi, M. 1996. *Integrating virtual reality functionality with traditional design tools.* DOE Research Contract 39/3/193, Department of Surveying, University of Salford, Manchester, U.K.

Furnham, A. 1998. *Personality and social behaviour.* New York: Arnold.

Hadipriono, F.C. & Larew, R.E. 1996. Safety training in virtual construction environment. *Proceedings of the 1st international conference of CIB working commission W99/Lisbon/Portugal, 4–7 September.*

Hale, A.R. & Hale, M. 1972. *A review of the industrial accident research literature.* London: *Her Majesty's Stationery Office.* ISBN: 0113608950.

Huang, G.Q. 1996. Developing design for X tools. *Design for X: concurrent engineering imperatives.* London: Chapman & Hall. ISBN: 0-412-78750-4.

Lingard, H. & Rowlinson, S. 1991. Safety in Hong Kong's construction industry. *The Hong Kong Engineer* 19: 38–44.

Lingard, H. & Rowlinson, S. 1997. Behavior-based safety management in Hong Kong's construction industry. *Journal of Safety Research* 28(4): 243–256.

Ng, K.G. 1992. *Reusable components for business information systems.* Dissertation for the Master Degree of Science in Data Processing, University of Ulster.

Prasad, B. 1996. *Concurrent engineering fundamentals Volume 1.* New Jersey: Prentice Hall PTR.

Petersen, D. 1984. *Human-error reduction and safety management.* New York: Aloray Inc. ISBN: 0-913690-09-0.

Sametinger, J. 1997. *Software engineering with reusable components.* Berlin: Springer. ISBN 3-540-62695-6.

Sush, N.P., Bell, A.C. & Gossard, D.C. 1978. On an axiomatic approach to manufacturing and manufacturing systems. *ASME Journal of Engineering for Industry* 100(2).

APPENDIX A

1. Use of the model to conduct a risk assessment on a Hong Kong Housing Authority residential block.

2. The user is observing the project from the ground level.

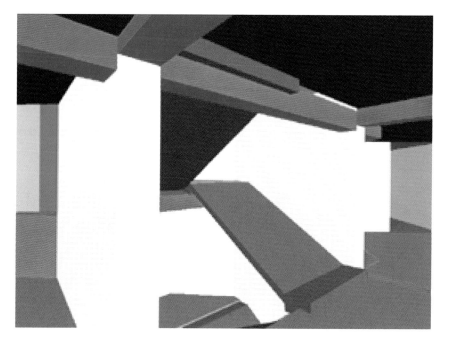

3. The user is on the lift area at core.

4. At the installation of wall reinforcement bars.

THE POTENTIAL OF 4D CAD AS A TOOL FOR CONSTRUCTION MANAGEMENT

Robert M. Webb[1], Theo C. Haupt[2]

[1] *Bovis Lend Lease, Charlotte, NC, USA*
[2] *M.E. Rinker, Sr. School of Building Construction,*
University of Florida, Gainesville, FL, USA

Abstract

4D CAD presents several opportunities for use as a tool for construction management with respect to the way that it links the temporal and physical spatial aspects of a construction project. Bovis Lend Lease has used 4D CAD to graphically represent the relationship between space and project schedule, through the actual transformation of that space over time during the construction of a building or facility. When it is considered that construction managers are constantly on the look out for effective ways to gain competitive edge in a highly competitive industry, 4D CAD has the potential to provide such an edge. 4D CAD provides the vehicle by means of which it is possible to integrate the functions, roles, responsibilities and relationships of, and between, all the participants in the construction process. This process is examined in this paper. Some of the problems, which need to be overcome, to make 4D CAD more attractive for construction management are also explored.

Keywords: 4D CAD, construction management, visualization, computer simulation

INTRODUCTION

Construction managers are constantly being bombarded by the need to make rapid and informed decisions in order to satisfy the traditional project parameters of time, cost and quality. Decisions need to facilitate completion within the project schedule and within the project budget while satisfying the desired quality requirements. As the number of technological options increases, so does the complexity and the cost of choosing which combination of available options is the most appropriate for a given application. Informed decisions involve the management of vast amounts of information about the combinations of available options and the simulation of their performance (Papamichael, 1999). To further exacerbate matters, the industry has become more complex due to several factors that include the greater use of specialist contractors, more off-site manufacture and assembly and the increased use of bespoke systems (Marsh & Finch, 1999). Additionally, manual methods are becoming increasingly difficult to implement at comprehensive levels. Consequently, decisions are made that are partially informed, resulting in missed opportunities, and unaccountable, undesired effects (Papamichael, 1999). These consequences are undesirable in the context of the increasing competitive environment of the construction industry. The rapid advances in information technologies and the continuously decreasing cost of computing power present promising opportunities for the development of computer-based tools that may significantly improve decision-making.

The combination of the graphic potential of 3D CAD with the construction project schedule is commonly known as 4D CAD. The 4D CAD technology presents several opportunities for use as a tool for construction management with respect to the way that it links the temporal and physical aspects of a construction project. It graphically represents the relationship between space and project schedule through the actual transformation of that space over time during the construction of a building or facility. Techniques that are presently being used to manage the design, planning and construction processes of a building facility, abstract the processes to produce a Gantt chart or CPM schedule (McKinney et al., 1996). A more comprehensive tool that will simulate and visualize construction activity sequences as part of an interactive experience is preferable. The interactive 4D CAD model provides just such a tool, in terms of which design and construction planning alternatives and decisions are evaluated, optimized and justified within the context of space and time.

Bovis Lend Lease has used 4D CAD prominently in the marketing, procurement and preconstruction phases of their construction operations, where it is primarily used as a visualization tool to help best plan a construction project (Fig. 1). Bovis Lend Lease has only recently begun using it in the actual construction phase. In this paper, these 4D CAD applications by Bovis Lend Lease are referred to, as well as a few others, with reference to their impact on construction management.

ANTICIPATED GAINS

The most obvious anticipated gains from the use of 4D CAD are that it will give the entire construction project team of clients, design consultants and contractors, a more effective tool to:

- improve communication between them while facilitating informed decision-making;
- facilitate the evaluation, implementation and monitoring of design changes;

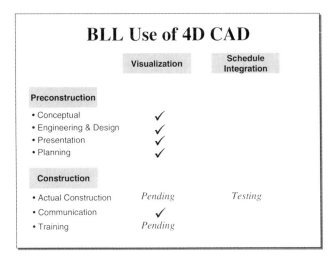

Figure 1.   Bovis Lend Lease use of 4D CAD.

- evaluate alternative materials or other processes to be used in the facility being planned or being built;
- evaluate and develop the most effective material staging and handling procedures for the project;
- identify and develop alternatives when disruption to the original plan on the construction project occurs;
- effectively train, and communicate with, construction crews (as well as other government regulatory organizations, community groups, etc.) specially before engaging in an intricate, challenging, or hazardous activity or a new construction method or technique;
- monitor progress on the project by comparing as-planned with as-built;
- improve the use of just-in-time material deliveries which are particularly important on construction sites where space is at a minimum or premium; and
- help overcome language barriers for members of the construction team, especially in the context of international construction activities.

However, to be able to meet the challenges that these applications present, 4D CAD as a tool must be capable of producing interactive 4D models for 4D animation (McKinney et al., 1996).

CONSTRUCTION SCHEDULES

4D CAD enhances the communication, approval and improvement of construction schedules by various parties, such as construction managers, clients, designers,

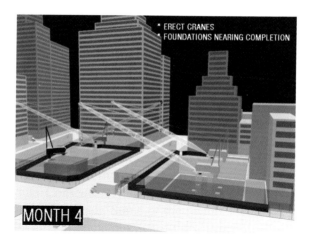

Figure 2.   Month 4 of the project schedule.

Unfortunately, most of the available simulation programs make this objective difficult to achieve. They were originally developed by researchers, for research purposes, and are not easy to use (Papamichael, 1999). They require significant amounts of detailed information about the building and its context, are very expensive to use due to the time required for the preparation of input and inter-pretation of the consequent output.

Bovis Lend Lease used an early generation of a 4D CAD visualization tool on its Lynchburg General Hospital project in the early 1990s. This hospital was preparing to undergo a renovation and addition to its existing facility. The visuali-zation tool proved to be very effective in helping the client and the entire project team plan and understand the sequencing of the work in such a way as to cause the least amount of disruption, and yet complete the project at the lowest possible cost.

More recently, Bovis Lend Lease has used 4D CAD visualization to facilitate the communication and planning for several projects in New York City. In an area like New York City where staging and logistics are particularly challenging, this application has been particularly helpful since nothing was left to the imagination in the visual presentation. The following graphics illustrate how it was used on one of the engagements.

By clicking on the date in the construction schedule, the construction activities that are scheduled or planned to be in operation will be graphically represented in 3D. In actual fact, Figure 2 illustrates the erection of cranes in month 4 of the pro-ject schedule. At this time the foundations were nearing completion.

Figure 3 shows month 15 of the project schedule indicating the expected pro-ject progress at that stage. Operations evident in this particular graphic, were the steel top out, setting boilers and cooling towers as well as con-ed permanent power. Month 22 of the project schedule is highlighted in Figure 4, where the operations included the removal of sidewalk bridges and the removal of the hoists.

Figure 3.   Month 15 of the project schedule.

Figure 4.   Month 22 of the project schedule.

It is clearly evident from this example how effectively 4D CAD can capture, and dynamically manage, the interaction between project components and resources over time, visualize these interactions, and support the real-time inter-action of users with the 4D model. With this tool, it is possible to enhance the communication, approval and improvement of construction schedules by various participants in the construction process, such as construction managers, clients, designers, sub-contractors and community members.

It is also possible to demonstrate the environmental impacts of the proposed project in order to allay fears, and gather support for it.

Another one of the popular software packages on the market that is used for 4D CAD visualization allows the project team members of a project to streamline

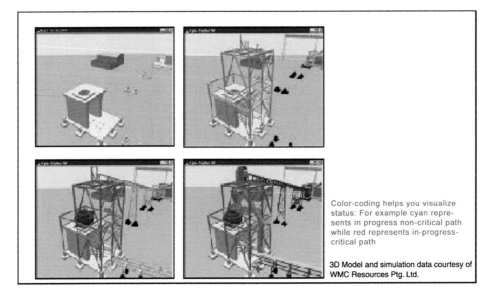

Color-coding helps you visualize status: For example cyan represents in progress non-critical path while red represents in-progress-critical path

3D Model and simulation data courtesy of WMC Resources Ptg. Ltd.

Figure 5.   Parallel workflow.

parallel workflow (Fig. 5). In a typical project environment, detailed engineering and scheduling run in two simultaneous yet independent work processes. This application integrates the design and construction planning disciplines, improving constructability and shortening the time from project concept to project completion.

Overall, this particular 4D CAD application can potentially provide the following benefits:

- definition of the scope of projects at project proposal and conceptual design stages;
- early development of a construction and commissioning or start-up method statement;
- involvement of construction and commissioning personnel in conceptual design stage;
- full investigation of design, constructability and commissioning issues prior to commitment of costly site resources;
- improved design and procurement strategies;
- better focus on pre-fabrication, pre-assembly and just-in-time procurement;
- smooth materials management and handling; and
- exploration of alternative dispute resolutions.

Bovis Lend Lease used a fairly specialized 4D CAD visualization tool on a project in Sydney, Australia. On this 50-story tower, there was a need to keep the reinforced concrete frame as light and open as possible while still being adequate for wind and other general structural requirements. The Strand 7 software tool was

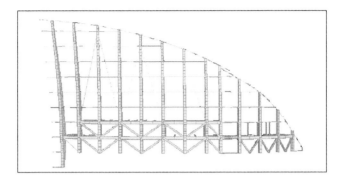

Figure 6.   Bovis Lend Lease project in Australia.

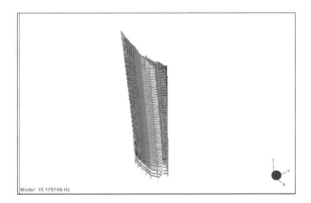

Figure 7.   Bovis Lend Lease project in Australia.

used to model the core structure. Wind was then introduced in time intervals to test for flaws in the design. Figures 6 and 7 illustrate some of the individual sequences from this visualization tool. This process added significant value resulting in a structurally sound building being constructed without the traditional amount of reinforcement that might have been required. The open, airy appearance of the building added to its aesthetic. The addition of open space throughout the structure was an added benefit.

ILLUSTRATION OF 4D CAD WITH SCHEDULE INTEGRATION

Bovis Lend Lease is actively using 4D CAD with schedule integration on a mixed use office park in London. This engagement has the essential ingredients for allowing 4D CAD to make a meaningful contribution; a very supportive team comprising of owner, design team, construction manager and key trade contractors. The initial objectives of the project team for this effort were as follows:

- to graphically represent planned progress element by element;
- to integrate graphical representation in time terms with current planning systems;
- to involve trade contractors to a greater degree in the planning process by giving them the opportunity to test their individual programs on the model;
- to have greater discipline in the process with both planners and trade contractors;
- to provide a tool to graphically test on-site program scenarios and recovery programs; and
- to provide historical records with respect to progress achieved and the reasons for that progress.

The following elements were built into the 3D models for all three buildings making up the project:

- all foundations for the concrete columns, steel core and external steel columns;
- all concrete columns for the undercroft to the third floor level;
- all floor slabs;
- steel core;
- external glazing;
- external staircases, steel bracing and external metal sun shade louvres;
- roof mounted mechanical plant;
- all external glazing;
- raised floors;
- wall construction around the steel core; and
- buildings to be positioned according to the site plan.

Figures 8 to 10 illustrate several specific times during the sequencing of one of the office buildings in the project.
Figures 11 to 13 are another illustration of project sequence on the office project in London.

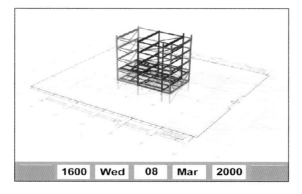

Figure 8.   Bovis Lend Lease project in the United Kingdom.

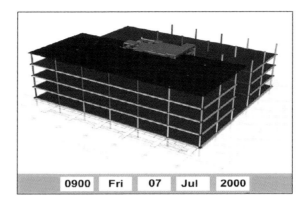

Figure 9.   Bovis Lend Lease project in United Kingdom.

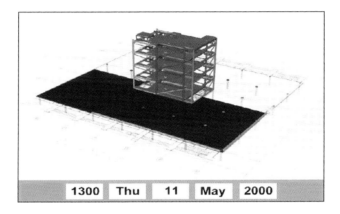

Figure 10.   Bovis Lend Lease project in United Kingdom.

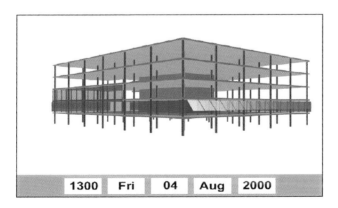

Figure 11.   Bovis Lend Lease project in the United Kingdom.

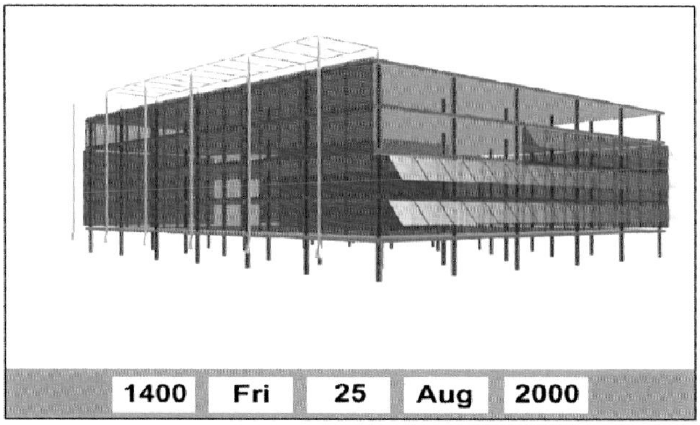

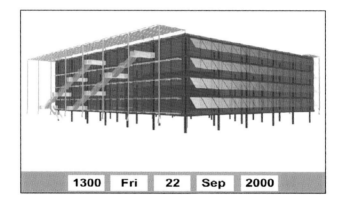

Figure 12.   Bovis Lend Lease project in the United Kingdom.

Figure 13.   Bovis Lend Lease project in the United Kingdom.

LESSONS LEARNED FROM 4D CAD WITH SCHEDULE INTEGRATION

The time required to build a model
The problem of design complexity versus programming time is one that poses the greatest risk to the successful development of the 4D CAD system being ready in time for use on a project. The current models are in a CAD platform and then animated, allowing them to operate a significant level of detail. This level of detail, however, carries with it a lengthy programming period.

The fact that computers can handle complexity does not mean that there is no need to design for simplicity. The model does not necessarily have to be as powerful as the CAD program. A simple stick model which can be built and operated by a contractor operative may be more usable and adaptable to change than a cumbersome all singing, all dancing model. Intricate details are not necessary on a 4D CAD tool. These details can be viewed adequately on detailed design drawings from either the trades or consultants.

Simplification of the model
With this point in mind, a simplification of the system currently being developed by the architect, sacrificing detail for usability, may offer a product that is more functional than the current envisaged format. The adaptation of the model currently being developed by the architect is a result of a hybrid of ideas from multiple sources, i.e. contractor, owner, consultants and architect. Indeed, simplification arises from knowing the job intimately; from a stand-back perspective or even from an innocent look at the project from the eyes of an outsider.

Alternatively, the current system of modeling might be used on projects where the design is more complete and perhaps more basic. The design and build form of contract where the contractor heavily influences the design, may have more merit than a Contract Management contract where the design is still evolving when construction has often started.

Project team summary observation
Given the complexity involved in building a 4D CAD model with the level of detail and chronological animation used on this project, impracticalities might occur in the sheer amount of time required to produce such a model. As design is very rarely 100% complete prior to construction, it is unrealistic to expect the design to be sufficiently evolved early enough before construction to allow a detailed model to be built. Perhaps this product should be viewed initially as a problem-solving tool for complex interface areas on projects. Then, by following the natural evolution of a product in regular use, the ease and speed of operation should increase and the role could be expanded.

OPPORTUNITY FOR CONSTRUCTION PROCESS IMPROVEMENT

The construction industry reputedly still suffers from one of the highest waste factors of all industries. It has been estimated that 25% of building costs in the United States are due to waste (CIOB, 1994). The introduction of 4D CAD technology into construction management could help reduce construction waste significantly. If the use of 4D CAD tools could reduce the cost of waste by 10%, the savings to the US $380 billion construction industry would be approximately US $38 billion, enough incentive to actively pursue 4D CAD or other efforts to reduce this waste factor.

The Construction Industry Institute (CII) found, in a specific study of industrial projects, that the average cost of rework on industrial projects exceeded 12% (CIOB, 1994). For the projects studied, deviation costs averaged 12.4% of the total installed project cost. However, the same study revealed that not all of the deviations on a project were recorded. For example, construction changes made at the site were often not included in format reports. It was not unusual for errors to be made good and/or accepted immediately rather than expending the time and effort to file formal requests. The deviation data gathered included only the direct cost of the rework for the item in question and included no indication of impact on the rest of the project. It is therefore concluded that both the number and costs of deviations reported for the projects in this study are conservative estimates of the actual values. The two major categories resulting in deviations were design and construction. By managing and tracking design and construction changes effectively using 4D CAD, a fair proportion of the costs of construction waste can be reduced and even eliminated.

CAD SOFTWARE SELECTION CONSIDERATIONS

According to a study conducted by Horne et al. (1999) in the United Kingdom, the following key independent variables for CAD software selection criteria were identified:

- modeling capabilities with respect to accuracy, ease of use and surface characteristics;
- visualization capabilities with respect to high quality, animated images.

In the same study, the dependent variables identified included:

- credibility of the data;
- comparability with other representational methods;
- appropriateness to differing needs of interested parties such as, for example, clients, designers, constructors, and sub-contractors;
- reliability;
- communicability;
- practicality in terms of time and other resources;
- reproducibility; and
- generic applicability.

The selection of the most appropriate visualization and modeling software is problematic because of the many features in different combinations, and incomparable user interfaces. Different simulation programs use different representations of buildings and their context, depending on the performance aspects that they address.

CONCLUSION

While 4D CAD has potential for improving construction management, providing competitive edge, reducing construction waste and costs, contractors have not yet accepted this potential on a large scale, especially in the area of construction management. The 4D CAD tool empowers construction designers, schedulers, superintendents, and project managers to develop a project schedule directly related to a 3D model of the building facility while at the same time, facilitating the understanding of the spatial and temporal relationships involved. Furthermore, 4D CAD technology will need to be available at a more affordable cost to enable it to be applied to jobs other than large, complex projects, both with respect to physical enormity as well as dollars.

Additionally, consideration has to be given to how the geometry of architectural form and structural design produced by architects and engineers can more easily form the basis for 3D representation while being linked at the same time to a construction schedule on a fully integrated basis. This aspect is essential if 4D CAD is to become the effective tool that it has potential to be with respect to construction project management.

Interoperability of software is essential for the continued development and deployment of 4D CAD. Central to this effort is the work being done by the International Alliance for Interoperability (IAI). This group is supporting standards that allow objects to transfer seamlessly from one application to the next. Additionally, the aecXML effort is quite important to the continued development of 4D CAD. This technology provides an effective, cross-platform, cross-application transfer of defined information objects.

Other areas of concern include the process of performance evaluation, the complexity of design information with respect to matching design and context parameters that are in conflict, and information overload caused by each decision being dependent on a large number of other decisions. Essential to the continued evolution and benefits of 4D CAD is the improvement of the environment in which it is to be used. 4D CAD needs to be promoted at all levels of the industry, including owners, contractors, designers, consultants, trade contractors. By having more projects online doing electronic collaboration will likely also increase the potential use of 4D CAD.

While the potential benefits of 4D CAD for construction management are not in dispute, the challenge faces software designers to produce an integrated system which can effectively address the concerns raised on a cost- and time-effective basis. Collaborative efforts across the various building construction related disciplines are necessary to realize the overall vision of a computerized building industry. The ideal with respect to construction management would be multiple simulation tools and multiple databases that are all interoperable in a distributed, networked environment between all participants in the construction process and beyond, to the eventual end-user.

REFERENCES

Horne, M., Hill, R. & Giddings, R. 1999. Visualization of photovoltaic clad buildings. *Building Research and Information* 27(2): 96–108.

Liu, L.Y. 1996. ACPSS—Animated construction process simulation system. *Computing in civil engineering; Proceedings of the third congress, Anaheim, California, 17–19 June: 397–403.* New York: American Society of Civil Engineers.

Marsh, L.E. & Finch, E.F. 1999. Using portable data files in the construction supply chain. *Building Research and Information* 27(3): 127–139.

McKinney, K., Kim, J., Fischer, M. & Howard, C. 1996. Interactive 4D CAD. *Computing in civil engineering; Proceedings of the third congress, Anaheim, California, 17–19 June: 383–389.* New York: American Society of Civil Engineers.

Papamichael, K. 1999. Application of information technologies in building design decisions. *Building Research and Information* 27(1): 20–34.

Riley, D.R. 1998. 4D space planning specification development for construction work spaces. *Computing in civil engineering; Proceedings of international computing congress, Boston, Massachusetts, 18–21 October: 354–363.* Virginia: American Society of Civil Engineers.

The Chartered Institute of Building (CIOB) 1994. *Constructing total quality handbook*: 7. Berkshire: CIOB.

VIRTUAL REALITY: A SOLUTION TO SEAMLESS TECHNOLOGY INTEGRATION IN THE AEC INDUSTRY?

Raja R.A. Issa

*M.E. Rinker, Sr. School of Building Construction,
University of Florida, Gainesville, FL, USA*

Abstract

A construction project is often divided into work packages because of its complexity. Although the construction of a big project becomes easier in a specialized industry, it also brings difficulty to the communication and cooperation between the participants of the project. It is proposed that a computer-integrated system may reduce the downsides brought by the fragmentation of the construction industry and improve the productivity and efficiency of the construction project. Different models of integration have been suggested, however, a uniformly accepted integration system has not yet been defined. The introduction of virtual reality (VR) technology into integration research may provide a general solution to this dilemma. A VR platform supported by knowledge-based database systems can become the main interface to construction information for every specialty throughout the construction (life) cycle of the project. All major application packages would be developed under or integrated in the VR system. As a consequence, we can foresee a marked decrease in legal disputes among the owner, architect, and constructor because of misinterpretation of design drawings and specifications and unmet owner expectations.

Keywords: construction industry, integrated construction environment (ICE), project modeling and integration, immersive, non-immersive, virtual reality

INTRODUCTION

In order for virtual reality (VR) applications to be successfully implemented in a complex industry such as construction, they must be part of a vertically integrated construction environment (ICE). Whether immersive or non-immersive techniques are used in the VR applications, users must be able to visualize design and construction information in 3D, photo-realistic, and interactive images. The user must also be able to interact with external applications at real-time, thus, allowing VR systems not only to be used as presentation tools, but also as a universal interface for all construction applications. Finally, construction professionals must be able to view, alter, test, etc. any function or part of the proposed design and at any stage of the project life cycle through the virtual space.

Due to the magnitude and complexity of construction projects, the traditional way of doing business in the construction industry is to divide the whole project into work packages according to well-established specialization. The work packages are assigned to specialty designers and contractors respectively. Although a system like this brings significant benefit to the industry, it also results in difficulties in communication and it requires extensive collaboration among the participants of the project.

The communication between the segments of the project relies mostly on drawings and specifications. Project participants acquire from these paper-based media information only relevant to their own specialty. Confusions and delays often occur due to the abstract nature of the said media and the process of constant reinterpretation by the project participants. Although computer applications in every specialty benefit the industry very much, most of these applications can only keep information integrity inside their specific areas. The communications between these independent systems are very limited and sometimes frustrating at best.

An established concept, Computer Integrated Construction (CIC) may provide a solution to this dilemma. Teicholz & Fischer (1994) defined CIC as "a business process that links the project participants in a facility project into a collaborative team through all phases of a project". The process included in this concept covers the whole duration of the project from design, construction to facility management. The main purpose of CIC is to facilitate information exchanges and collaborative efforts among the project participants. A summary of the objective of CIC was given by Teicholz & Fischer (1994) as: (a) rapid production of high-quality design, (b) fast and cost-effective construction of facility, (c) effective Facility Management.

VTT (1998), the Technical Research Center of Finland, proposed the interesting analogy of the current integration research in the construction area as shown in Figure 1. The independent computer applications in specific areas like design, construction and project management, which shows the fragmentation of the construction project, was referred to as "Islands of Automation" or "Islands of Information". The contour line is actually the time axle. The current coastline means the frontier of the research and applications at present, while the coastline of 2000 was the goals that the researchers may achieve before the next century. With the advances of the computer technology, breakthrough of some key concepts, and the effort of both researchers and industry practitioners, "the water level has dropped" (Froese, 1994), and bridges are built between the islands. This process will eventually lead to a "unified continent", an integrated construction management (CM) system. Figure 1 is an imaginative description of the evolving process of integrated computer applications in construction industry.

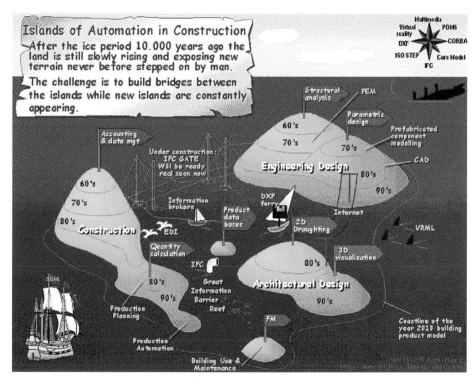

Figure 1.   The islands of Information (from VTT, the Technical Research Center of Finland).

Table 1.   The development of integration.

Time	Architectural design	Engineering design	Construction
1960s	–	Structural design	Accounting, data management
1970s	–	Parametric component design	–
1980s	2D drawing	CAD in drawing, prefabricated component modeling	Quantity calculation, production planning
2000 and beyond	ISO STEP, VRML, Internet, EDI, DXF, Information Broker …		

The VTT analog can be converted into a tabular format, as shown in Table 1. It provides a historical perspective for the general scenario. The first use of computer in engineering design was in the area of structural design. Although some sophisticated structural analysis theories were developed long before that time, they could not be implemented without powerful computers. The computerization of accounting systems that happened in the 1960s was the first full-scale acceptance of computer application in business management. At the same time, computers also began to be used in construction process.

The development of computer hardware advanced in the 1980s to enable higher graphic processing ability. CAD, which stands for "Computer Aided Design", became popular. However, it is more like "Computer Aided Drawing" in most cases. The software that was used to generate drawings for the construction or manufacturing process had difficulty exchanging information with other software, such as structural analysis or other computation software. Estimating software, which is closely related to accounting system, came into use in the 1980s. During this time the CPM method was computerized as well and became the mainstream of construction planning software.

The great advances in technology integration came in the 1990s. Our whole society was affected by the fast development of computer technology. CAD technology advanced from 2D drawings to 3D visualization and VR using VRML became possible. It was recognized that the lack of integration between computer applications in this area could become a major obstacle to the further development. Integration plans were proposed and tested in order to achieve full-fledged production automation under the control of a unified computer system.

The rapid expansion of the Internet has resulted in numerous possibilities and opportunities for the construction industry to make improvements to many aspects of its business operations. Some new areas of applications started emerging, such as product databases, and facility management archives. The current active areas of standardization research include:

- ISO STEP (Standard for the Exchange of Product Model Data): BCCM Core Model, Express, etc.;
- CORBA (Common Object Request Broker Architecture) from OMG (Object Management Group);
- IFC (Industry Foundation Classes) by IAI (International Alliance of Interoperability);
- PDMS (Product Data Management);
- Multimedia (Video and Audio), Internet, VR, and DXF.

These standards form the three important sub-areas of computer integration research:

- *Integration of the existing applications*: The major purpose is to establish information standard between computer applications developed independently. It is like the transportation facilities between the islands shown in Figure 1.
- *Investigation of new specialties*: Some applications are specialized in new specialty areas, such as product databases and facility management.
- *Introduction of new concepts and utilities from computer science*: The latest developments in computer science give rise to new possibilities for solutions to current problems in the construction industry. Examples of such developments include, VRML, Internet, XML, EDI, DXF, etc. These new developments appear as objects floating around the islands in Figure 1.

HISTORICAL PERSPECTIVE

The history of Computer Aided Project Management (CAPM) can be traced back to the 1950s. The ultimate aim of such a system at the time was toward creating an integrated management entity for the construction project. The actual developmental process did not go as predicted, however. Constraints from both technical and managerial aspects hindered further investigation. The pursuance of integrated systems began in the past decade again because of the new possibilities brought up by the following changes in construction industry and advances of computer technology.

CONSTRUCTION INDUSTRY NEEDS

The construction industry was long considered slow in adopting new technology. It "has viewed innovation with suspicions or attempted to protect new thinking by protectionism" (Brandon & Betts, 1997). While the manufacturing industries improved their productivity and quality by leaning production and applying worldwide manufacturing benchmarking studies of production standards, the construction industry remained low profile. Once the construction industry realized this, major companies in the industry along with research institutions began to investigate a solution that might bring profound innovation to the whole industry. This lead to the setup of an international network, Construct IT, which represents one of the most productive efforts ever made in the quest for integration in construction. A major purpose of this network is to promote the application of information technology, system integration and standardization in the construction industry. The development of these applications is a necessary step toward a fully integrated construction industry.

THE CHANGING STYLE OF PROJECT MANAGEMENT

The evolution of project delivery systems brought changes to the general style of project management. The basic project delivery system, which has a long history, is called the "traditional method". It can be described as a process of "design–bid–build". The owner, architect, and contractor are three independent parties bonded by contractual or administrative relations. Some new delivery systems, which were referred to as "alternative methods", emerged in the practice of the past decades and began to challenge the domination of the traditional method. These systems include: CM, design-build, and BOT.

A notable feature of the change is that the later systems tended to have more centralized management (Clough, 1986). The project management responsibilities were conveyed to another independent party in the CM method, the construction manager in construction stage. The project management team, consisting of the architects, contractor, and owner's representative, is headed by the construction manager. In the Design-Build method, a single contract including both design and construction responsibilities, is awarded to a "design-builder". The design-builder is responsible for controlling the project activities for the duration of the project. While in the BOT method, the financing, operation, and limited time ownership appear in the job tasks of the builder.

DEVELOPMENT OF COMPUTER TECHNOLOGY

The information produced from a construction project can be enormous because of the complexity and large scale of the construction project. It can be extremely hard to manage construction activities in an integrated manner without the help of computer facilities.

The major factors that influence the further progress of integration research include:

The computational capability of computers: Graphic and database application need the support of higher process ability.
Hardware cost: Sharp decreases in hardware prices make possible the expanded usage of computers in the construction industry.
The concept of databases: Orderly organized information provides efficiency and increases productivity.
Networking: Brought a revolution to the method of communication in construction, which is crucial to the cooperation and coordination in construction process.
AI and neural networks: Added further strength to the integrated construction system.

The emergence of new concepts and methodology like Object-Oriented languages and databases, and the Internet are providing even more possibilities for the construction industry to integrate its systems. Sometimes it is just a matter of using our imagination to discover new potentials of computer in construction domain.

CURRENT RESEARCH

The integration issues in construction were investigated intensively in Nordic countries, such as Sweden and Finland. Some of their publications are leading studies in this area. Their research actually defined the structure and trends of the integration research. The major institutions endeavoring in this area in North America include CIFE at Stanford University and the University of British Columbia.

VTT, the Technical Research Center of Finland, Finland
VTT, the Technical Research Center of Finland, is an expert organization that carries out technical and techno-economic research and development work. There are two research groups in VTT related to CIC, RATAS and Project Planning and Building Design group. The active researchers include Matti Hannus and Mika Lautanala.

Center for Integrated Facility Engineering, Stanford University
The Center for Integrated Facility Engineering was founded in 1988 as an industry affiliated program of the Departments of Civil Engineering and Computer Science within the School of Engineering at Stanford University. The center is working on applying information technologies to the construction industry to improve integration in the construction process from the design to the management of the constructed facility. The research involves a wide range of technical, social, economical and managerial issues. The major topics explored include: CAPM, the strategy issues of the CIC, and information exchange standard. CIFE has many publications and has made great contributions to the establishment of some basic concepts in this area. Key researchers include Paul M. Teicholz, Hans Bjornsson, Raymond Levitt, and Martin Fischer.

Department of Civil Engineering, University of British Columbia, Canada
The integration research is very impressive due to the efforts of T.M. Froese, who received his Ph.D. degree in Civil Engineering from Stanford University in 1992. One of their research interests is the design of integrated, computer-based decision tools to support project design and construction. Their major works include the development of improved tools for modeling projects, representing and selecting construction technologies, encoding construction expertise into systems, automating the interpretation of construction records, and capturing multi-media project information. Findings from this work have been put into practice on many construction projects. Other research within this area focuses on information sharing and the integration of project functions throughout the construction life cycle. A major methodology of their integration research is to introduce new concepts of computer science and technology, such as Object-Oriented database principles, into construction industry.

VIRTUAL REALITY: A SOLUTION TO INTEGRATION?

The basic concept of VR is to model the shape of the objects in three dimensions. The idea of VR appeared decades ago, but the inferior ability of the primitive computers at the time hindered data-intensive implementations. The price of equipment was so prohibitive that the application of VR had to stay in a virtual status. However, VR does have some advantages that put it among the most promising solutions to implement system integration.

The "ideal" solution

A VR Integrated Construction System can be expected to

- enable designers, developers, and contractors to use the VR system and virtually test a proposed project before construction actually begins;
- offer "walk through" view of the project so that problems can be found and design improvements can be made earlier;
- provide free flow of information between CAD systems and other applications work packages by professionals in industry, minimize the misinterpretation between participants in the project, especially between designers and clients;
- facilitate the selection of alternative designs by allowing different plans to be tested in the same virtual world.

In a VR Integrated Construction System, VR becomes the main interface for all application packages and construction information for every specialty throughout the construction (life) cycle of the project.

Two ways of interacting with a VR world

There are two approaches to implementing a VR World: immersive and non-immersive. In an immersive approach, the user is surrounded by the virtual world through curved screens and body suits or headmounted devices (HMD). The audio and visual perception of the user will form a virtual world. The non-immersive approach, also known as desktop VR, enables users to interact with the virtual world with conventional devices such as a keyboard, mouse and a monitor. Although this does not give the same level of spatial awareness as the immersive approach, it does provide users with a low cost solution and does not require the use of the HMD. This solution seems to be an attractive compromise for many users who are uncomfortable about spending a long time in a helmet (Issa, 1999).

VR can be interpreted as a bridge between subject and human perception. These two ways of implementing VR provide solutions from two ends of the bridge. The immersive approach makes human perception its focus, while the non-immersive approach started from the description of the subject. The distinctions between these two styles of VR may eventually diminish with technological advances. But for the current investigation of VR in construction, the non-immersive approach seems to be more applicable.

Problems of current VR systems

Currently the two major areas of functionality of VR in construction are interaction with objects in real-time and walk-through presentation. These features are mainly about visualization and simulation, instead of providing a basic interface between users and the project (subject).

Most of the time VR systems are just supplementary to CAD packages. They cannot perform standalone design let alone be the bases of 2D drawings and all engineering design. Lots of implementation problems come from the supplementary role of VR systems, and include difficulty in use, requirement of special skills, and expensive to implement. These problems, which mainly come from the lack of integration between application packages, constitute tremendous barriers to the implementation of VR systems in the real world.

What is needed to make it happen?
To make the dream of VR come true, a scheme similar to the following needs to be set up:

- VR must become the general interface among the different applications instead of their individual interface.
- 2D and 3D images must become not just a way of presentation, but more importantly they must become interface for interactivity.
- A central core which is a database system (most likely a knowledge-based database system) will be the basis of the whole VR system, the application and the interface.
- The VR integrated construction system must be able to reside on a communications network (the Internet or more precisely the WWW).

A serious challenge to the actual deployment of a VR system is whether an Industry Standard is developed or not. Before a complete solution can be provided to the user, the industry must be persuaded to adapt and move to a totally new, standards-based system.

VR APPLICATION PROTOTYPES

Construction material specification integration
The integration of construction drawings, design and material specifications within a VR environment allows the AEC professionals and the owner/procurer of construction services to preview the final product of their effort. This preview allows the participants in the project to more realistically determine the soundness of the design; the appropriateness of the construction techniques and the adequacy of the facility and material finishes in meeting the owners needs, prior to the execution of the project. Consequently, the expectations of the parties will be more realistic and the risk of costly disputes will be reduced considerably.

Collaborative virtual prototyping

Even though VR-based tools can be useful at every stage of the construction process (to convince clients, to design the project, to organize and follow the construction site, etc.) important applications are related to the "design phase". Decisions taken during the early design phase are of paramount importance due to their possibly dramatic effects on the final project, timing and costs. Virtual prototyping allows architects, engineers, contractors, and clients to create a design and evaluate it simultaneously for function, cost and aesthetics very early in the design process.

The visual capabilities and the interactive inspection features offered by VR-based tools are much more extensive than those offered by standard CAD tools. Furthermore, coupled with distributed technologies such as STEP and CORBA, VR tools offer cooperative capabilities very useful in the design, by geographically distant teams, of large engineering projects. In that case, the virtual prototype can be considered as the starting point of the design process. After the first stage where the design teams test and validate the virtual prototype, relevant data is extracted from this prototype and is fed into CAD/CAM tools in order to be completed with more technical and detailed data (Fig. 2).

Link with CAD tools

The reverse process (i.e. extract data from CAD/CAM tools in order to visualize objects in a VR tool) is also possible. Nevertheless, it requires a fair amount of simplification (for evident reasons of performance optimization, detailed data cannot be

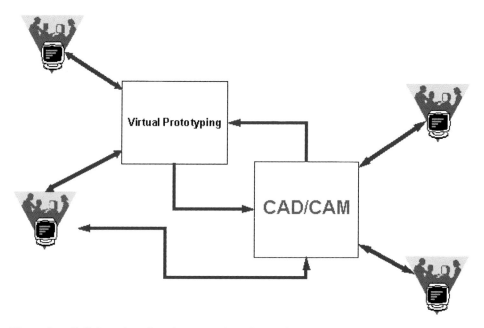

Figure 2.   Collaborative virtual prototyping (CIB, 1999).

fed into VR tools as a whole). Furthermore, existing techniques of simplification (polygonal reduction, re-meshing, etc.) still have some limitations particularly for granularity management (a small component that highly effects the virtual scene, e.g. a key hole when simulating lighting effects in a dark room, might be suppressed in an automatic re-meshing procedure). CAD models aim to represent the geometry of components for their manufacturing or for executing physical simulations (deformations, thermal analysis, etc.) by using methods such as the Finite Elements Method.

On the other hand, VR models aim to represent objects visually, in order to interact with them.

CAD models, therefore, can only be used within VR platforms after being processed by optimization procedures such as tessellation. Tessellation can be described as the processing of a 3D model in order to reduce the number of triangles of the model while maintaining an acceptable visual aspect. This procedure has some limitations:

- it could *change the frontiers* on the components of the initial model (which might be a problem when, for instance, two components should keep a perfect fit);
- it is rather limited in handling *gaps* and *intersections* in the model.

In both cases, manual corrections are usually needed to rectify the simplified model before using it in a VR application. Furthermore importing CAD models within VR tools usually yields a model where some of the facets are missing. This is due to the fact that, in CAD tools, a common way of constructing 3D models is based on *symmetry* (i.e. only half of the model is described and the other half is deduced by using symmetry axis). The "symmetrical copy" of the 3D model would be identical to the original one but would have *inverted normals*. When imported into a VR platform that uses *backface culling* for optimization issues, the "symmetrical copy" of the model will not be visible. A manual action from the user is then needed in order to invert the normals of the model. An interesting optimization tool for CAD/VR coupling is CAD-Real-Time Link from Prosolvia Clarus (http://www.clarus.se).

VR applications for detailed design
During the *detailed design phase,* virtual prototyping tools will allow the design office to refine the design proposed by the architect by adding constraints and modifications induced by the technical calculations (structural, thermal, lighting, etc.):

- *Acoustics*: The results of acoustic calculations can be related to the sound going through a window or a wall or the sound inside a room (e.g. a meeting room). These results are usually 3D sound WAV files associated to the related building components.
- *Lighting*: Different lighting calculation methods can be used. The most effective ones are based on *radiosity* computation and *raytrace* rendering.

These methods combined give a high realistic visual feedback on the architectural options taken.

- *Thermal analysis*: At this stage, thermal analysis is done in order to estimate the performance of HVAC systems and/or the comfort in the built environment. This should give a quick feedback on the architectural options taken (orientation, glazed surface, etc.).
- *Documentation/annotations*: During the design, users should be able to access, in line, to relevant documentation and standards about the building components. This can be done by supporting hypertext links between building components and related URLs. Furthermore, users can attach annotations to a given component or the overall project so they can leave a message or explain a choice to other users (that are not in the same work session).

Construction projects can very easily become complex. Therefore, performance optimization procedures are of paramount importance. Two optimization procedures are particularly efficient in the AEC sector: *scene graph culling* (when the walkthrough takes place in the first floor, there is no point in loading the geometry of the other floors) and Levels of Details (LOD) (each of building components, that can be very complex if represented with all these details, have several representations that will be displayed depending on the LOD required based on the distant of the component from the camera). These methods, combined with more generic optimization methods (such as *visibility culling* and *backface culling*) should allow complete scalability of the system regardless of the complexity of the construction project.

THE INTEGRATED CONSTRUCTION ENVIRONMENT

In order for VR applications to be successfully implemented in a complex industry such as construction, they must be part of an ICE (Fig. 3). In such an environment, construction applications packages are integrated through a central intelligent core whereby project information is controlled, maintained, and manipulated. The user interface for this environment should have the ability to convey project information in a humanly acceptable level, i.e. elements, spaces, resources, etc. At this end, VR can play a major role in the development of a human computer interface for the ICE. Whether immersive or non-immersive techniques are used, users can visualize design and construction information in 3D, photo-realistic, and interactive images. The latter facility allows users to interact with external applications at real-time, thus, allowing VR systems not only to be used as presentation tools but as a universal interface for all construction applications. Construction professionals can view, alter, test, etc. any function or part of the proposed design and at any stage of the project life cycle through the virtual space.

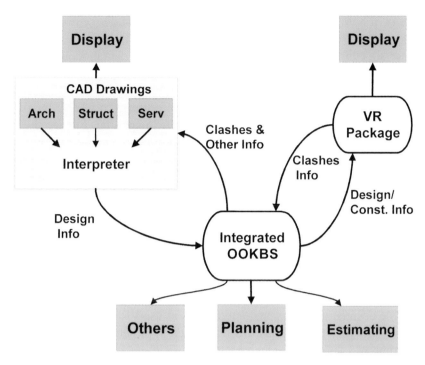

Figure 3.   Conceptual presentation of the ICE.

In the short term, VR (non-immersive) can be used, as a modeling tool, to complement current design tools such as CAD systems. This implies that VR can be considered as an application package within the ICE, which aims at providing flexible, realistic, and interactive presentations. Once VR models are generated in virtual space, users can navigate through the product, at its current stage of development, and interact with any design elements or spaces to access further information or run external simulation programs. Users' movements and queries are monitored and controlled by the intelligent central core of the ICE.

VR, as a universal interface, can be enhanced by video conferencing. Communica-tions between different members of the design team or between design team, builder and owners can be significantly improved by using a combination of VR and video conferencing techniques. If VR models are generated automatically from the traditional design tools at the local design office, such models can be transmitted to the client's remote site. Clients can navigate through the product and/or request alterations to the design or part of the design by simply pointing or moving the concerned elements. Alternative solutions can then be suggested by the designers and represented to the clients for final approval. The same scenario can be applied to improve communications between various members of the design team.

A prototype of such product has already been developed by the Automation and Integration in Construction (AIC) research group at the TIME Research Institute,

University of Salford. At its current stage of development, the prototype "SPACE" (Simultaneous Prototyping for An integrated Construction Environment) integrates six construction applications with the central data models. The applications are: design, specifications, estimating, construction planning, site layout planning, and VR (Alshawi & Budeiri, 1993).

In the long term, VR (fully immersive) will offer the average user the potential to enhance the final presentation by combining 3D images, headmounted displays, sounds, and self-movements. The ability to support the illusion of the individual's movement through the virtual space will make the implementation of VR much more acceptable to humans. Users will be able to feel/see their movements in space, thus, improving the performance and well-being of the ultimate human user. Users' movements and requests, in virtual space, will be monitored and controlled by an intelligent and integrated knowledge-based system and other external construction applications where all communications with external applications' are carried out in virtual space in either a textual or graphical format.

The flexibility offered by virtual environments to visualize and interact with the virtual world, provided that these technologies are available at a reasonable cost, will enable designers, clients, and contractors to use VR to rapidly construct and test their prototypes before constructing the actual project. But this only happens if the strengths of the technology are emphasized and the hype is significantly played down. VR should be treated not as a technology in its own right, but in terms of a suite of technologies, which when carefully implemented, are capable of matching the capabilities of humans to the requirements of the application or task he or she is required to work with.

The potential of VR can only be realized if it is integrated with construction applications packages. An ICE should be developed where all construction applications are integrated through a central intelligent core. VR can play a major role in the development of a human computer interface for such an environment. Whether immersive or non-immersive techniques are used, users can visualize design and construction information in 3D, photo-realistic, and interactive images. Moreover, VR displays and interactive devices should only be selected on the basis of (a) human factors issues, i.e. what is expected of the performance and well-being of the ultimate human user, and (b) customer requirements.

ROBOTICS INTEGRATION IN THE CONSTRUCTION WORKFORCE THROUGH VR

Mobility on the job site

Robots in construction are part of a system made up, as shown in Figure 4, of four basic, interacting components: operator, computer, robot, and the construction environment. The design of new robots to supplement the construction workforce

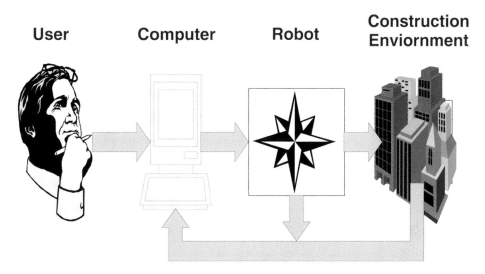

Figure 4.   Robotic system model.

can only be achieved with the help of VR. VR can be of valuable assistance in both geometric aspects, such as link dimensions, work envelope and dexterity, as well as in control aspects, such as, visualizing sensor data and virtual navigation controllers. By combining and integrating reflex control and virtual environments, great progress can be made toward completely autonomous robots.

Reflex control allows us to establish a direct link between information and action, thus bypassing the high resource overhead requirements associated with the decision making stage. This inclusion of decision in information is only possible in well-identified environments (Burdea & Coiffet, 1994).

Virtual environments and fixtures
Applying VR to unstructured environments involves two categories of virtual objects. The first category would involve the modeling of the minimal information known about the unstructured work environment. The result of the process will be the replacement of an unknown characteristic of the real environment by a known virtual environment. This principle could be extended to most characteristics of the unstructured construction environment.

The second category of virtual objects is "virtual" fixtures or guides, which help during the task execution. Lines, curves surfaces, or volumes of known geometry, along which the robot is restricted to move (Coiffet, 1993). Figure 5 shows a conceptual representation of a system involving virtual environments and fixtures.

Teleoperation
Another approach to dealing with the unstructured construction environment is by keeping an operator involved in the control loop. Using VR to integrate the

Real Robot   **Virtual Fixtures**   **Virtual Environment**   **Real Construction Enviornment**

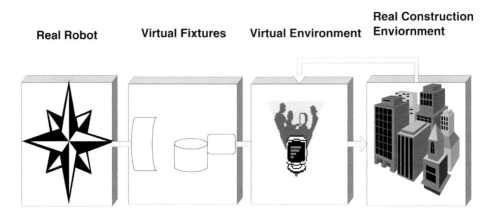

Figure 5.   Virtual environments and fixtures concept.

operating environment with the operator's environment, facilitates maneuvering the robot on the unstructured environment of the construction job site (Rosenberg, 1992). One of the difficulties involved in dealing with teleoperation is that the operator is remote from the robot and the feedback data may be time delayed or insufficiently detailed for correct control decisions.

Stereo viewing
Human vision is the most powerful sensorial channel and has extremely large processing bandwidth. Our depth perception is associated with stereopsis, in which both eyes register an image and the brain uses the horizontal shift in image registered by the two eyes to measure depth (Julesz, 1971). Depth perception is what allows us to maneuver in our environments, because it gives us our ability to see scenes in 3D. Integrating robots in the construction workforce and in the work environment will involve designing a display and vision system that can adequately provide this type of stereo vision and allow for its integration and interpretation in terms of maneuverability.

Dexterity in manual functions
Once the autonomous robot has reached its designated work area on the construction job site, the focus shifts from mobility to dexterity in performing construction tasks. Dexterity training for robots can be achieved by fitting a human construction worker with an instrumented glove and then asking that worker to perform tasks identical to those expected from a robot in the future. The system, as shown in Figure 6, uses a pair of electrodes placed on the forearm, which are connected to a neural network computer. Once the neural network has been trained, the robotic arm can be used to perform the function it has been trained on with relative dexterity.

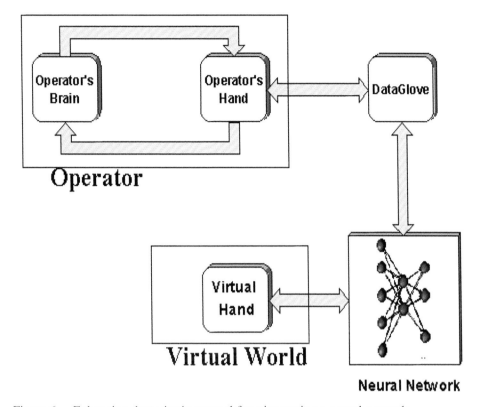

Figure 6.   Enhancing dexterity in manual functions using a neural network.

CONCLUSION

Research into CIC has just begun to draw the attention of both industry and academic institutions. Non-immersive applications using VR as a modeling tool to verify the integrity and constructibility of designs have been gaining in popularity. At the same time collaborative design-build efforts are making the use of VR as a universal interface among construction team members ever more popular.

   With the ever-increasing computational power available to users, it is expected that VR technologies and peripherals will develop rapidly and their application will have the potential to change dramatically the way of doing business in construction industry. What is not certain is what path the advances will take and what kind of impact these advances will bring to the industry. Using the VTT analogy presented in Figure 1, we can reasonably conclude that the ground under the water is still not clear, but is getting clearer.

REFERENCES

Alshawi, M. & Budeiri, M. 1993. Graphical simulation of construction sequence by integrating CAD and planning packages. *The International Journal of Construction Information Technology* 1(2): 35–46.

Brandon, P. & Betts, M. 1997. *Creating a framework for IT in construction.* The Armathwaite Initiative, the Formation of a Global Construction IT Network, Construct IT Centre of Excellence.

Burdea, G. & Coiffet, P. 1994. *Virtual reality technology.* New York: John Wiley & Sons.

Clough, R.H. 1986. *Construction contracting,* 5th edition. New York: John Wiley & Sons.

Coiffet, P. 1993. *Robot Habilas and Robot Sapiens.* Paris: Editions Hermes.

Froese, T. 1994. Information standards in the AEC industry. *Canadian Civil Engineer* 11(6).

Issa, R.R.A. (ed.) 1999. *State of the art report: virtual reality in construction.* International Council for Building Research Studies Documentation (unpublished).

Julesz, B. 1971. *Foundations of cyclopean perception.* Chicago: University of Chicago Press.

Rosenberg, L. 1992. *The use of virtual fixtures as perceptual overlays to enhance operator performance in remote environments.* Technical Report, Center for Design Research, Stanford University, September.

Teicholz, P. & Fischer, M. 1994. Strategy for computer integrated construction technology. *Journal of Construction and Management Engineering* 120(1): 117–131. ASCE.

VTT, the Technical Research Center of Finland 1999. http://www.vtt.fi/cic/ratas/islands.html

CONSTRUCTION MANAGEMENT PULL FOR nD CAD

Peter Barrett

University of Salford, Salford, UK

Abstract

4D CAD work at present could be typified as "techno-construction-centric". This paper endeavors to provide a wider construction management perspective that will open up high value alternative areas for consideration. Construction is a dynamic, fragmented and combative industry. There is just not a stable platform for the adoption of sophisticated tools. In addition it is usual to speak of managing for time, cost and quality which is really quite misleading. The following performance dimensions are suggested: location (planning), function, aesthetics, cost, time, health and safety and environmental performance. This implies a broader, longer-term perspective beyond immediate project needs.

Given the *tacit–tacit* emphasis of the industry, the mismatch with the *explicit–explicit* character of 4D CAD systems is stark. Instead of accuracy and detail, coarse robustness and connectedness are needed in systems that cover the important hard and soft dimensions. The implication is that 4D CAD systems need to shift emphasis towards the *tacit–explicit* mode by accommodating the above wide range of hard and soft, long- and short-term performance dimensions (nD CAD). In parallel with this a push towards supporting *explicit–tacit* knowledge conversion is needed with an emphasis on richer communications. The developments suggested will create a closer fit between the characteristics of the systems and the reality experienced by those in the industry. As such it will simply make more sense for such systems to be taken up through industry pull.

Keywords: construction management, knowledge transfer, nD CAD, tacit knowledge, industry pull

INTRODUCTION: CURRENT 4D CAD FOCUS

To date 4D CAD appears to be focused on integrating the technical design information respectively within the design and construction phases (e.g. Aalami &

Fischer, 1998). This has great potential to unlock the synergies between the knowledge and experience of the designers and that of the constructors. These islands of know-how are typically isolated by education, tradition, orientation, contracts and processes. The thrust of the work, however, seems to be limited to the technological issues, with a heavy emphasis on the construction phase. In short, 4D CAD work at present could be typified as "techno-construction-centric". This paper endeavors to provide a wider construction management perspective that will open up high value alternative areas for consideration. In this way, it is hoped, the full benefits of the emerging technology can be developed. Unless otherwise stated a UK perspective is being taken.

Implicit in this paper is the view that very seldom is the technology itself the area where fundamental problems in practice arise, there is usually some better or worse technical solution. Limiting underlying assumptions, however, the innovation processes necessary for take-up and the management of people are much more difficult to handle. The usual "best practice" solution is to advise that people should behave more rationally (Barrett & Stanley, 1999), but this belies the reality of human nature. In practice what has to happen is that the nature of construction players is accepted as a given (at least in the short to medium term) and that initiatives, such as 4D CAD are developed taking this into account.

Direct experience of using shared CAD systems to integrate project planning on a large trial project in Norway (studied in Barrett & Stave, 1993) reinforces this view. One manager closely involved stated: "The most important basis of success in the introduction and development of new technology is not the technology itself, but the people who are to use it." It was how the technology supported changing perceptions, relationships, information flows and working systems that mattered, not so much the particular form of the technology chosen. Thus the remainder of this paper will seek to elaborate on some of the wider factors and opportunities that any 4D CAD system should ideally attempt to address and support.

THE NATURE OF CONSTRUCTION

Construction is a dynamic, fragmented and combative industry, certainly in the UK, and it would seem worldwide (e.g. Latham, 1994; DETR, 1998). The ability to absorb technologies is hindered by the industry having its own unique "recipe" of assumptions, knowledge-bases, technologies and practices (Huff, 1982; Spender, 1989). This "recipe" considerably erodes the ease with which technologies can be transferred into an industry by creating "incompatibility barriers." These barriers generally can only be surmounted by the technologies being carefully interpreted and transformed to blend comfortably with, and enhance, the recipient industry's "recipe". It has to be said that the construction industry is highly reactive and action orientated. This can suit coping with short-term emergencies, but is problematic when more reflective initiatives are needed.

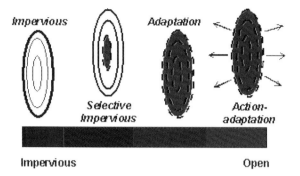

Figure 1. Impervious construction!

From the evidence of the industry's reaction to, say, quality, health and safety and environmental imperatives (Barrett & Sexton, 1999), there is a tendency to do the minimum, as late as possible. Drawing from Leavitt et al. (1973: 306–310), Figure 1 provides a continuum, from companies having impervious boundaries with their business environments to companies with open boundaries.

Very often construction companies are at the impervious extreme. Sometimes they are "selectively impervious", taking on some proposals where they seem to fit (or are unavoidable!), but grafting them on so that over time the company develops an array of incompatible systems that do not deliver synergies. In fact "initiative fatigue" is more likely. A few firms will organically adapt to their environment reactively, but hardly any will actively manage their business environment ("action-adaptation") for symbiotic benefit. The economic turbulence of the industry is one undoubted cause of this inability to deal with major change in a positive way.

Work by Sarshar et al. (1999) has articulated the problem in a way that will be familiar to IT specialists. Taking the Capability Maturity Model for the software industry, developed at Carnegie Mellon University for the US Department of Defense (SEI, 1994), she has worked with colleagues and industry to develop a version for construction companies. Apart from finding that the supply chain aspect is under-represented in the original model, they have found that the systems of even very good companies in construction appear to only be at Stage 1 or maybe 2 (Fig. 2), i.e. they are "chaotic" or moving towards "repeatable". In fact quite a lot of time on this industry-collaborative project was spent debating whether to create a Stage 0! Thus, the "organizational readiness" (Hersey et al., 1996) for taking up 4D CAD is likely to be rather low. There is just not a stable platform for the adoption of sophisticated tools, particularly if they are prescriptive in the way work is done.

IMPLICATIONS OF TURBULENCE AND IMMATURITY

"The construction industry is very old, but not very mature" (Barrett & Sexton, 1999). It does not, and arguably should not, have a high level of standardization or

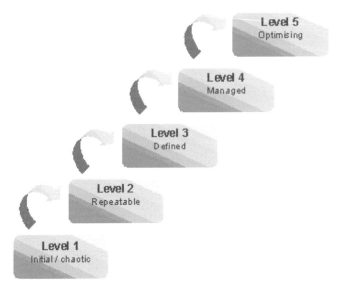

Figure 2.   CMM/SPICE maturity levels.

a large body of explicit knowledge. However, there is a massive accumulation of fractured, formal, façade systems dating from as far back as the Middle Ages. Construction needs to be recognized as a "new" industry in which an emphasis on innovation, customization and the use of tacit knowledge is celebrated and supported (Hansen et al., 1999).

Cairncross (1998) has graphically illustrated by extrapolation "how the communications revolution will change our lives" in general and the OECD (1999) has highlighted the effect electronic commerce will have on the time dimension in particular. It is reasonable to assume that rapid change will occur within construction in the coming decade. More powerful, cheaper computing power will be available to more computer-literate workers. This must be used to bind back together the industry by supporting strong informal horizontal linkages as well as formal vertical integration. Galbraith's (1977) model given in Figure 3 further illustrates this suggested emphasis.

His model is proposed as a complete set of alternatives to absorb "exceptions" in a company, i.e. gaps between work demands and worker capabilities.

Traditionally the construction industry has relied on formal (contract) mechanisms (1, 2 and 3) together with self-containment (5), witness the prevalent division between designers and constructors, and, of course alternative 4—"slack resources". This last is a euphemism for "sub-optimal performance". However, new technology will undoubtedly support an increased capacity to process information. This may be aimed at creating vertical information systems (6), but as a complementary approach, or indeed, an alternative, supporting the creation of

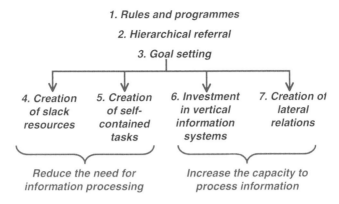

Figure 3. An information processing view of the firm.

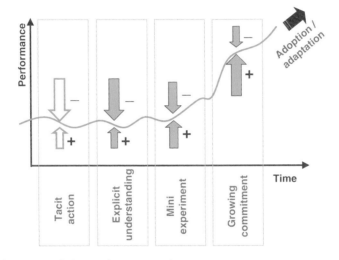

Figure 4. Incremental change in construction.

stronger "lateral relations" (7) has much to commend it. This approach concerns horizontal communications. The importance of this aspect will be further drawn out in this paper. These changes will need a "fusion" of technological and organizational innovations (OECD, 1998).

Work on the implementation of change in construction has highlighted the need for an incremental approach that emphasizes the adaptation of the technology to the company for success (Barrett & Stanley, 1999). Figure 4 shows the process revealed when attempts were made by the author to implement consensus improvements to the briefing process. The collaborating companies started from a position of doing what they always had because it had worked so far, that is their actions were based on a tacit knowledge base. At this stage there were only minor irritations such as "driving forces" and quite significant "restraining forces" (Lewin, 1947).

First of all, as we tried to implement the agreed changes together, nothing happened! The firms decided their clients "didn't want to be guinea pigs" and carried on as before. Even though the researchers were disappointed we had to try to understand this resistance. It became clear that it was based, not so much on antagonism towards the proposals, which they had all been involved in developing, but rather it was grounded in a highly rational (from their perspective) aversion to risk. Each firm was not acting in a vacuum, they had many relationships with other parts of the industry. If they changed in isolation it could cause real problems of disjuncture in these relationships. Given the current nature of the industry, any resulting problems would be blamed on our partners. In these conditions a company would have to be either foolish or brave and highly motivated to move first. Initially our companies were not sufficiently motivated, but they did see more clearly what was going wrong and why. This led to the explicit understanding stage shown on the model, i.e. no change in their actions, but a significant change in the companies' appreciation of the impact of their actions, even as they rushed from incident to incident.

As a consequence of this heightened awareness the companies tended to gain motivation to change "now that we have seen it go wrong again!" This increased motivation led to the design of low risk "mini-experiments" with consequently reduced restraining forces. The forms were beginning to move. As these experiments delivered some benefits the companies commitment to the implementation of the ideas grew. The perception of the risks dwindled and the motivation to carry on grew. Further experiments were built in and the ideas progressively adapted and adopted by the companies. This rather extended description serves to highlight the rocky road to implementation and so the need for a sustained incremental approach and a tolerance (indeed expectation) of a mixture of success and failure on the way. This rather uneven progress is reflected very well in the juggling analogy (Gelb & Buzan, 1994), where it is stressed that to make progress at all mistakes must be accommodated. In litigious construction this means trying things out behind the scenes or on a limited basis. Trying to pick up three balls in front of an audience and just juggle without any practice is likely to have comical results. However, not many construction practitioners want reputations as clowns! So take-up of new systems must allow adoption in stages through non-threatening, low risk, incremental access.

There is very strong resonance between the above view of managing change within construction companies and the findings of MIT's recent study of the processes of architectural design. This is summed up in the following view: "No process worth replicating is replicable. Put in less jarring terms: A worthwhile process must be reinvented rather than mechanically reproduced." (Horgen et al., 1999: 269). The point is that there is no short cut in the adoption of new technologies and integrating them with a company's processes. Each firm has to engage in its own tailored learning process. If technology is at the center of this it is still only a part of a much more complex whole. And beyond the firm itself and all of its dimensions there is the industry context to take into account. A firm may choose to move, but on its own it cannot move very far.

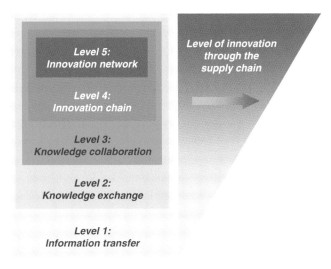

Figure 5.   Innovation and the supply chain.

The briefing study quoted above also stressed the high-leverage potential for improvements that create and maintain a shared vision amongst all involved in a construction project. This has been reinforced and extended in a study of construction innovation which stressed that to make significant gains a strategic approach, utilizing a carefully selected portfolio of company-to-company relationships, is needed (Barrett & Sexton, 1998). Various possible levels of interaction are set out in Figure 5. In these collaborative endeavors, soft factors such as trust and longer-term plans were found to be central, reflecting Doz's (1996) formulation of a developing cycle in strategic alliances against the three criteria of efficiency, equity and flexibility. Doz makes the telling point that "the impact of initial conditions quickly fades away" in successful alliances. Again this argues against top-down imposition of the ready-made, complete, "right" solution.

For 4D CAD this seems to argue for flexible shell systems that support a good deal of integrating features to reflect the reality that diverse players will be using them. Trying to make everyone play the game by a single set of rules is not likely to work. Providing an environment that a company can incrementally take-up could. Providing better means of flexible communication could.

For an example of this latter aspect, during case studies of the operation of supply networks for hybrid concrete systems (Barrett, 1998), it became very apparent that the formal system of controls could not cope with the volume and rapid changes of information. However, the communications technologies supporting the "informal system of controls" (Tavistock, 1966) that took over, such as radios, facsimile machines and mobile telephones, were very widely used with no resistance, no hesitation. They met the needs of the workers and fitted with the culture of the industry (if 4D CAD systems can do the same then there will be no problem in achieving take-up!).

The above observation links well with the analysis by Hansen et al. (1999) of knowledge management in management consultancies. They make the distinction between two principal strategies, namely "codification" or "personalization". The aim with the codification approach is to "Provide high quality, reliable, and fast information-systems implementation by reusing codified knowledge". Personalization aims to "Provide creative, analytically rigorous advice on high-level strategic problems by channeling individual expertise". (p. 109). Codification assumes a volume of work with a lot of reuse of knowledge by large teams with a lot of juniors, using people-to-document systems, highly supported by IT. Personalization assumes high margin work carried out by well-qualified small teams using person-to-person knowledge management, supported by moderate IT systems.

> … companies that use knowledge effectively pursue one strategy predominantly and use the second strategy to support the first. We think of this as an 80-20 split … Executives who try to excel at both strategies risk failing at both. (p. 112).

So, they are arguing you have to make a broad choice, but which alternative relates best to construction? To choose a predominant strategy (codification or personalization respectively) depends on whether: the service is standardized or customized; the organization/sector mature or innovative; the knowledge used to solve problems explicit or tacit knowledge. Much of construction is customized and we have already seen that company systems are of low maturity and the knowledge used is predominantly tacit. This all points towards a strategy that emphasizes teams, person-to-person knowledge management and only moderate IT support with an emphasis on communications.

It is not possible to generalize, but it seems that construction faces the uncertainty and, doubtless as a consequence, has many of the characteristics of a new, dynamic, thrusting industry. These characteristics should be celebrated and supported. The work of Hansen et al. connects with more substantial work specifically focused on construction professionals. For example, Coxe et al. (1987) set out a parallel choice between "practice-centered businesses" and "business-centered practices". The great majority of design firms are primarily practices and this again fits with the customization strategy.

This emphasis on communications rather than codification underpins a speculative view based on work with the UK concrete industry. This is drawn from case study work specific to concrete, but also from an innovation study investigating supply chains in which the prominent role of materials and components suppliers became very evident. Figure 6 sets out the idea of creating a web-based environment to link the various players. In particular, design support is made accessible to designers and tendering and ordering made easy for contractors and suppliers. Within the shared environment a set of broad generic systems are provided with associated interactive cost, etc. models. The starting position is shown in the diagram, but over time it is anticipated that consortia and market mechanisms

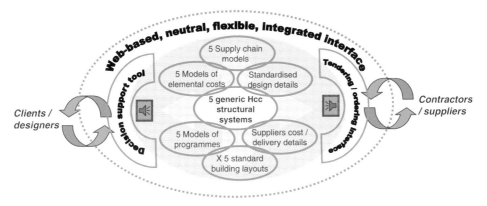

Figure 6.   A speculative, neutral web-based environment for concrete.

Table 1.   Employment and number of construction enterprises in the EU.

Size of firm (employees)	Number of firms	Percentage of firms	Number of employees	Percentage of employees	Percentage of total turnover
0–9	1,700,797	92.80	3,512,969	43.3	36.1
10–19	76,618	4.20	1,025,263	12.7	12.4
20–99	48,695	2.70	1,820,354	22.5	24.7
100–199	3,543	0.20	492,320	6.1	7.2
200–499	1,585	0.10	483,257	6.0	7.5
500+	585	0.03	761,345	9.4	12.1
All firms	1,831,822	100.00	8,095,509	100.0	100.0

would evolve the content of the shared space as experience and opportunities became evident.

The above examples generally relate to studies of large organizations with relatively robust in-house management capabilities. So, it is worth mentioning in passing that a series of benchmarking studies by ConstructIT (1998) of large construction companies has found that there is great variability in the beneficial use of IT systems even amongst these leading companies. However, the message that non-prescriptive, flexible communications tools are needed is heavily reinforced when the size structure of the industry is considered. Construction is populated mainly by very small firms. Although drawing from slightly old data, in the EU for example, about 97% of firms, doing 49% of the work are less than 20 employees strong (Atkins, 1994; see Table 1 for more detail).

This means that sophisticated bespoke technologies are unlikely to be adopted by enough players to create sufficient momentum for an industry-wide, top-down change. This has led construction researchers in the UK towards using standard software platforms where possible, or at least making the interface appear familiar. For example, a database tool used to encourage cross-organizational learning in construction (COLA) is available in Microsoft Access, as "smaller organizations are more likely to have access to the ... products and, more importantly, access to the skills required to adapt and manipulate the database for their own purposes" (Boam, 1999).

DIMENSIONS OF SYSTEMS

It is usual to speak in project management circles of managing for time, cost and quality. This is really quite a misleading formulation. For example, time and cost can easily be seen as quality dimensions themselves, and, what else is to be included under the quality heading anyway? It seems more fruitful to think in terms of generic performance criteria, or at least to use quality as an overall descriptor for a comprehensive set of such criteria. The former "quality-free" approach is advocated convincingly by Sjoholt (1989), however, the latter approach was taken in a project involving the then newly independent states of Estonia and Lithuania, together with Danish and UK partners (CONQUEST, 1995). In this challenging cross-cultural project we strived for an objective assessment of quality in construction. The following list of eight performance dimensions was created: location (planning), function (fitness for user's purpose), aesthetics, cost, time, technical performance, health and safety and environmental performance. These criteria were found to be "owned" variously by different selections of stakeholders, ranging from the "paying client" to "society at large", with any control achieved by a range of mechanisms, ranging from socialization through custom, to legislation.

The resulting mapping achieved is summarized in Table 2. Solid squares are the primary mechanisms and unfilled squares the secondary mechanisms. This table was developed for a particular comparative analysis, however, the blank sheet stimulus provided by the new states studied helped the team take an objective view of construction in general. The notion of a range of performance dimensions controlled (managed) using a range of mechanisms in a variety of combinations seems robust.

This incidentally fits with work on the interactions of markets, hierarchies and networks, which Bradach & Eccles (1991) typify as underpinned by price, authority and trust respectively. They claim that these mechanisms are: "overlapping, embedded, intertwined, juxtaposed and nested ... typically control mechanisms are grafted on to and leveraged off existing social structures". This seems a realistic view from our studies of construction. The implication is again that top-down "designed" solutions are unlikely to be capable of reflecting the complexity of reality.

Taking this wider range of, say eight, criteria should help to overcome "over-measurement" of the easily measurable (Etzioni, 1964). It can be seen to lead naturally to the idea of a broader perspective beyond the immediate project needs. Construction can in fact be seen to be merely a change agent for the built environment, which itself supports society's needs (see Fig. 7). That is, although construction

Table 2. Performance criteria and their stakeholders.

	Quality criteria (What) by country							
	1 Location	2 Cost	3 Time	4 Aesthetics	5 Technical	6 Fitness for	7 Environment	8 Health and safety
Controlled by (Who)	Estonia · Denmark · Lithuania · UK	Estonia · Denmark · Lithuania · UK	Estonia · Denmark · Lithuania · UK	Estonia · Denmark · Lithuania · UK	Estonia · Denmark · Lithuania · UK	Estonia · Denmark · Lithuania · UK	Estonia · Denmark · Lithuania · UK	Estonia · Denmark · Lithuania · UK

Brief stage

1. Market
2. Custom
3. Legislation
4. Institution
5. Experts
6. Clients

Design stage

1. Market
2. Custom
3. Legislation
4. Institution
5. Experts
6. Clients

Construction stage

1. Market
2. Custom
3. Legislation
4. Institution
5. Experts
6. Clients

Post construction stage

1. Market
2. Custom
3. Legislation
4. Institution
5. Experts
6. Clients

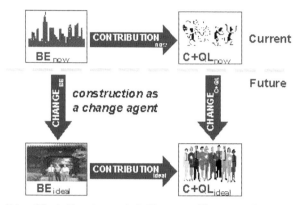

Notes: BE = built environment; C+QL = competitiveness and quality of life

Figure 7.   Construction as a means to a means to an end.

is a very big and important industry, it is a service industry and as such is a means (to a means) to an end, and not an end in itself.

The built environment is there to serve society, which in the UK's Government's terms equates to improving competitiveness and quality of life. Construction is the change agent through which this change is achieved. Thus, at a minimum the whole project life cycle needs to be addressed and here the "4" in 4D CAD comes into focus as time is a key linkage to many of these issues. This takes into account a long-term perspective as well as the project duration highlights issues, such as the whole life cycle of facilities, including user views, but also societal factors such as planning and environmental impacts.

IMPLICATIONS OF MULTI-DIMENSIONALITY

Creating systems that can model, in detail, physical project attributes over time, perhaps even including the variation of construction project costs, is clever, but does it address the critical problem areas? Systems like these will probably support improved efficiency in design and logistics on site, however, these are not the problems highlighted in strategic reports on the industry (e.g. Latham, 1994; DETR, 1998). In these reports there is a heavy emphasis on trust and relationships, as well as some fairly mechanistic thinking. This lag is evident in construction supply chain theories to emphasise almost exclusively logistics (O'Brien, 1997), whereas, in the general field of logistics there is an interesting development towards studying the softer and longer-term aspects of supply networks, such as perceptions of requirements and performance and customer satisfaction (Harland, 1996). This appears to be following the burgeoning services literature, but particularly Grönroos' (1984) work, which highlighted the "expectation—perception gap".

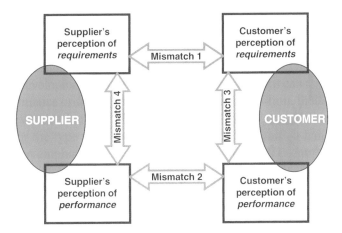

Figure 8.   Harland's supply chain model.

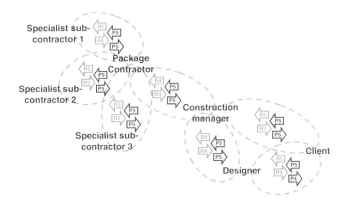

Figure 9.   Example analysis of supply network.

The complex "gaps" picture given in Figure 8 is, of course, a great simplification because each "customer" is someone else's "supplier". But, more than this, supply chains are really supply *networks* as shown in Figure 9. This was based on a study of a specific project. The difficulty of handling complex, "soft", interacting data like this is a big challenge for a 4D CAD system. However, it is a necessary effort if relevant, useful systems are to be created. Interestingly, this move in emphasis is paralleled in general UK research funding which is shifting away from technical "single loop" research (Argyris & Schon, 1978), on how to do better what we already do, towards supporting organizational and sociological research on how effectiveness can be enhanced through customer-orientated, knowledge-based innovations (e.g. HMSO, 1998). The UK Government is pushing forward on open electronic communications in its regulatory role (Cabinet Office, 1999), and this could facilitate and stimulate linkages right the way through from land survey data, to design, to construction, to

CLOSURE

R.R.A. Issa, I. Flood, W.J. O'Brien

M.E. Rinker, Sr. School of Building Construction,
University of Florida, Gainesville, FL, USA

CURRENT STATE OF 4D CAD

3D/4D CAD tools are in their second decade of use in the design and construction industry. From scattered early efforts in the late 1980s, the 1990s saw a proliferation of such tools in practice. Certain industries such as industrial construction now routinely design projects in 3D. In other areas, such as general building construction, there are owners, architects, contractors, subcontractors and vendors that have taken the lead in applying 3D/4D tools across their operations. Owners require 3D for design communication and decision-making and use the models for facilities management. Architects use 3D models to communicate with clients, render dramatic sculptural forms, and share these models with contractors and subcontractors to enable construction of these sculptural forms. Contractors use 3D models for coordination of building systems and materials procurement. Contractors use 4D models to ensure schedules are buildable and for trade coordination. Subcontractors use 3D and 4D tools for much the same purposes as contractors, but often use more detailed models to plan field production. Many vendors have developed 3D models of their offerings to allow architects and engineers to place them directly into their designs. Vendors also use 3D models to direct their internal production processes.

The 1990s also saw increased sophistication in software packages. At the high end, tools have grown more powerful in storing intelligence about building objects and their relationships (e.g. an object is a beam connected to a column; the beam has attributes of strength, material properties, a manufacturer and tracking number, etc.). An infrastructure of tools has been developed to support implementation of 3D/4D models. These tools include: libraries, global positioning systems for surveying linked to the 3D model objects, portable display devices, and data exchange standards. Collectively, these developments make possible the integrated application of 4D analysis across all stages of the project lifecycle (although an integrated off-the-shelf package does not yet exist, requiring investment on the part of project team members). Advanced software applications promising more integrated functionality are in the commercial product development pipeline.

Mid- and lower-level software applications have also seen increasing sophistication in their ability to manipulate 3D models. Mainstream commercial CAD packages now allow users to rapidly develop 3D models at different levels of detail. It is also possible to perform 4D analysis by linking the 3D models to scheduling packages and/or by manipulating the 3D objects using layer functions. 3D technology has also enhanced lower-end CAD systems, perhaps best evinced by the sub-US$100 home design CAD packages sold in retail stores. These systems allow homeowners to rapidly build and render 3D models of their homes, producing drawings usable by contractors. Cost and capability are no longer considered a barrier for any firm that wishes to employ 3D/4D tools.

Despite the power and availability of 3D/4D tools, their use is still not common in most areas of design and construction. Apart from modeling of piping and related systems in large-scale industrial construction, the most common use of 3D models is in marketing and conceptual design. Clients are sold on the building concept in 3D "walk-thrus." Sometimes these 3D concept models have a 4D element to portray the impact of construction on existing sites or to portray stages of project development. These models have little detail, and are seldom further developed through detailed design or for construction planning. This unfortunate circumstance is not solely the fault of uncreative practitioners. Further development in at least four technical and business areas is needed to fully realize the potential of 4D CAD on practice.

FUTURE STEPS—FOUR AREAS OF DEVELOPMENT TO TAKE 4D TOOLS TO THE NEXT LEVEL

1. *The visual interface*: There are two related problems with the interface of current tools. First, the more sophisticated tools (and many of the lower-end tools) are difficult to use. The models cannot easily be manipulated by anyone other than experts, and simpler representations (e.g. static models including printouts) lose much of the power of full models. Difficult-to-use software also limits the ability of users to add information to the model, consequently restricting the model's usefulness. Second, it is difficult to customize visualization of the 3D/4D models and related data. Every user will have a different set of needs for information and different preferences for visualization of that information. Even the most sophisticated of existing applications have limited abilities to display information in various forms. Hence, even if an integrated 4D model is available, the model may not support business decisions beyond its ability to display information. 4D tools

are by nature information rich. Improved ease of use in accessing and contributing information and greater fluidity in customizing the presentation of that information are necessary developments if 4D models are to be fully leveraged.

2. *Data exchange between applications*: Improvements in the visual interface require seamless exchange of data among the software applications behind the interface. Currently, code must be written for each link between applications. This is a lengthy and difficult set-up process that most projects are unwilling to support. While writing code on a per-project basis does allow customization, the approach is not scalable. Nor is the customization necessarily fluid; as project needs and project participants change, it is unclear that the code can be easily adapted to meet those changes. What is needed is theory and methodology about the sharing of information that supports implementation on projects without detailed coding by experts. Currently, standardized data models are under development to allow software applications to share data. These data models provide a basis for sharing data such as <schedule activity>. It is less clear that these data models will directly support higher level reasoning, particularly with regard to user level customization of information representation. Further developments in sharing and manipulating data are required for widespread use of 4D tools.

3. *Job design to leverage the tools*: While there does not currently exist a recipe for collaboration to make best use of 4D tools, it is clear from the existing implementations that the technology requires new ways of working together. These range from the simple changes of design review with a 3D model to the more sophisticated questions of who contributes what to the 3D/4D model. How job design and responsibilities should change is a fundamental issue with implications for firm and project organization, legal responsibilities, and, not least, contractual incentives. Many of the projects that have used 3D/4D models collaboratively across firms have done so in the spirit of experimentation. Thus, they have not addressed the issues of standard operating procedures using the new technologies. What these procedures should be is very much a subject for future research and development.

4. *Benefits and contracts for 4D*: Closely related to the idea of job and organization design is the appropriation of benefits using a 3D/4D environment. While many of the firms using 3D/4D tools report benefits in a wide variety of applications that they believe easily outweighs the cost of 3D/4D model development, only limited cost-benefit analysis has been performed. It is clear that early development of the model in the project lifecycle pays dividends later in the lifecycle. However, given the fragmented nature of the construction industry, many of the early developers of design are not responsible for later stages of the project and do not accrue the benefits from savings in those later stages. In any phase of the project, it is the rare firm that wishes to absorb the cost of model development without a clear understanding of the benefits and how they will be distributed. As it is difficult to predict where benefits will be realized (e.g. on one project in design coordination, on another project in productivity improvement), it is unclear how to

assemble contracts that appropriately reward those firms that bore the cost of developing a 3D/4D model.

The construction industry has a poor record of consistently supporting improved planning processes, and 3D/4D tools may not be different from other approaches. How to structure contracts to support the use of 3D/4D tools is unclear. Three approaches come to mind: first, the owner can simply pay for the models and assume that they will get their fair share of benefits. Some owners are taking this approach. Second, individuals in the industry can adopt the tools and create new forms of firms that simply do everything better than traditional firms. These "category killer" firms have redefined other industries and may do so in design and construction. Third, the tools can become so powerful and cost effective that they replace existing 2D tools and methods in the various firms involved in the project process. Thus, 3D/4D tools may organically replace 2D work practices. We have seen developments in each of these approaches. However, how these benefits are supported by contractual structures remains an open question.

So where do we stand on the use of 4D in construction? The 1990s saw the premature proclamation of a coming revolution in practice based on the early benefits and tools seen in the late 1980s. Now early in the first decade of the millennium, we are hesitant to make sweeping claims. But we do have the experience of the 1990s to guide us, and concrete benefits have been seen in practice. The contents of this workshop have established a roadmap to move forward, and the editors tentatively suggest that at the end of the next decade we will be writing not about novel developments in 4D tools, but about incremental improvements in a technology that is well accepted.

Subject index